Advances in the Protection of Museum Collections from Earthquake Damage

Advances in the Protection of Museum Collections from Earthquake Damage

Papers from a Symposium Held at the

J. Paul Getty Museum at the Villa on

May 3–4, 2006

Jerry Podany, *Volume Editor*

THE J. PAUL GETTY MUSEUM, LOS ANGELES

Published by the J. Paul Getty Museum, Los Angeles

Getty Publications
1200 Getty Center Drive, Suite 500
Los Angeles, California 90049-1682
www.getty.edu

Gregory M. Britton, *Publisher*
Mark Greenberg, *Editor in Chief*

Mollie Holtman, *Editor*
Sylvia Lord, *Copy Editor*
Elizabeth Zozom, *Production Coordinator*
Hespenheide Design, *Designer and Typesetter*

Printed and bound in China through Asia Pacific Offset, Inc.

Library of Congress Cataloging-in-Publication Data

Advances in the protection of museum collections from earthquake damage : papers from a symposium held at the J. Paul Getty Museum at the Villa on May 3–4, 2006 / Jerry Podany, volume editor.
p. cm.
Includes bibliographical references and index.
ISBN 978-0-89236-908-9 (pbk.)
1. Art—Conservation and restoration—Congresses. 2. Museum buildings—Earthquake effects—Congresses. 3. Earthquake resistant design—Congresses. I. Podany, Jerry. II. J. Paul Getty Museum.
N8560.A29 2008
069'.50289—dc22

2007047370

Front cover
Ceramic models representing ancient vessels being tested for stability on a shake table.
Photo: Department of Antiquities Conservation, J. Paul Getty Museum.

Contents

Foreword

Just two weeks before the conference that served as the source for this publication, the centennial of the Great San Francisco Earthquake of 1906 was observed. The terrible force that almost destroyed one of America's great cities, caused the loss of many of the city's beautiful monuments and buildings, and, most tragically, took so many lives, was again featured on the front pages of our newspapers and brought forward in our thoughts. While it would seem that nothing good could come of such a tragedy, much that we recognize of modern seismology and seismic mitigation has been realized in its aftermath.

It was appropriate then that the J. Paul Getty Museum gathered international experts, engineers, seismologists, and museum professionals to discuss the protection of cultural heritage in museums and collections in the anniversary year of that event—and within the first year of the opening of the renovated J. Paul Getty Museum at the Villa. The conference brought together experts from countries that are often shaken by earthquakes and whose collections are exhibited under the constant shadow of seismic threats. Over the two-day conference, these experts considered the advances that have been made in protecting collections, the needs that remain, and the benefits that future technologies might offer.

The J. Paul Getty Museum has accumulated some twenty-five years of experience in the area of seismic protection of its collections, and along with the Getty Conservation Institute, it takes very seriously the issues of mitigation and response, developing innovative approaches and continuing the search for ever more effective techniques.

It is important that we recognize the deeper significance of our efforts to protect our collections and our cultural resources from harm. It has been demonstrated many times that in the aftermath of a major disaster, those who have undergone the trauma seek out the very resources we are protecting. The public seeks comfort, beauty, and stability in cultural institutions, in history, and in art, and we have a responsibility to ensure that such resources are there when they are needed the most.

I am gratified by the response to the conference and by the resulting papers published here, as well as by the contribution this volume will make to worldwide efforts to keep museum collections safe from harm. It was an honor to welcome colleagues who came great distances—from Greece, Japan, Turkey, and India—as well as our neighbors from sister institutions in San Francisco and Los Angeles. I am pleased to present their contributions in this publication.

The conference took place over an enlightening and exciting two days, and these proceedings reflect the commitment, expertise, and creativity brought to the considerable task of keeping our objects safe when the very ground we stand upon seems to come alive and heaves with terrible force. I believe the increased understanding, collaboration, and knowledge that this congress has generated will, if we remain diligent, lead to a reduction of future devastation in the wake of these inevitable events.

Michael Brand
Director
The J. Paul Getty Museum

The Challenge

Jerry Podany

It will be helpful for the reader of this volume to consider the following equation: *acceptable risk = acceptable damage.* On one end of this equation is the desire to make collections accessible and to give them meaning through that accessibility—that's part of what museums do. On the other end is our desire to assure the long-term preservation of collections for future generations—something museums also do or, at least, should do. While there is a pragmatic limit as to how far we can minimize the second half of the equation, we continually strive to keep it as low as possible, even insisting on a balance of zero. After all, who among us would want to define what level of damage to our collections is acceptable? That insistence and commitment have given birth to a profession and an industry: conservation and collections care. Though fluctuating worldwide in the priority it is given, conservation has won the battle to convince others to recognize basic threats to the preservation of collections. Light, humidity, temperature, pollutants, and pests are now widely recognized as contributors to ultimate destruction. That recognition has brought many positive changes to the definition of a protective museum environment and has advanced appropriate collection-care practices as well as research in these areas.

But nature regularly unleashes larger threats, with catastrophic forces that take only minutes, even seconds, to destroy whole collections, buildings, and villages. The sheer power of floods, mud slides, fires, and earthquakes overwhelms us. Our lack of ability to predict the arrival of these events defeats the confidence we may have in our capacity to keep material heritage safe. In fact, however, we have at our disposal methods to prepare for these threats and ample techniques to mitigate their destructive results. Considering the availability of mitigating measures, one is left to wonder why more has not been done. Perhaps foremost among the reasons is that many museum and collections staff members are unaware of the techniques that can be applied to the problem. Certainly there is a lack of familiarity with the seismic engineering principles that lie behind many of the approaches that could be followed. Repeatedly we hear, "What can one do? If the earthquake is that large—well, everything will be lost anyway" or "Everything was fine in the last earthquake; I'm sure it will all be fine the next time." Such attitudes are only buoyed up by the apparent costs of effective measures and the lack of funding for preventative care. It is true, of course, that relative to the usual amounts budgeted for preventative care and relative to the overall budgets of most small museums, some mitigation techniques can be prohibitively expensive. But there are numerous simple and effective approaches to mitigation that carry minimal costs. Securing shelving can be done quite inexpensively, and moving big, heavy items from the top shelf to the bottom shelf requires only time, a little effort, and, perhaps most important, awareness.

Choices among possible methods and case studies exemplifying efforts already taken are the topics of this volume and were the focus of the conference that gave rise to this publication. The conference, "Protecting Collections from Earthquake Damage," which took place at the Getty Museum on May 3–4, 2006, came both at a moment of commemoration and at a moment of celebration. As Getty Museum Director Michael Brand has written in his introduction, the centennial of the Great San Francisco Earthquake was observed in 2006. The anniversary reminded us of the devastation that can be caused by earthquakes. But 2006 was also a time to celebrate the opening of the recently renovated Getty Villa in Malibu, which provided an opportunity to bring together experts involved in seismic mitigation to discuss positive steps forward. It was, after all, at the Villa site that the Getty's seismic mitigation efforts began, and where they have continued for more than two decades. It is interesting to note that the Getty Museum was the site of a much earlier conference on seismic mitigation in 1984. Among the attendees at that conference was Robert Rutherman, who had already authored a very early study on the protection of collections entitled "Reducing the Risk of Nonstructural Earthquake Damage: A Practical Guide." The 2006 conference was the result of a long-held dream of once again gathering experts dedicated to the neglected topic of seismic protection of collections. The meeting has resulted in the beginnings of new partnerships and long-term exchanges to widen interest in and encourage the development and application of seismic mitigation in museums around the world. To date, we can do nothing about the occurrence of earthquakes, except to continue efforts to increase our understanding of their character and cause. Perhaps someday we will have at our command the ability to accurately predict the moments of occurrence and the epicenters of earthquakes. No doubt such an ability would require far more complex statistical and geophysical analysis then we have today. For now, the suddenness of earthquakes and our lack of ability to specifically predict them sets earthquakes apart from other natural disasters, which at least present formidable signs of their possible arrival. We can, however, make a few solid predictions. We know, from empirical evidence, from historical records going back to antiquity, and from daily reports, that some regions of the earth are far more prone to earthquakes than others. And we know that it is never a matter of *if* but, instead, a matter of *when*. Those of us living and working in seismically active areas (a large proportion of the earth's population) should rely on a far more simple and direct statement of prediction—that is to say, *earthquakes will occur.* With that knowledge it seems unreasonable, even irresponsible, not to take more steps to combine expertise and efforts to mitigate the effects of earthquakes in order to reduce the vulnerability of our collections. While we should always encourage research in mitigation, we need not necessarily wait for it. Steps to reduce the destructive effects of earthquakes are available to us now. And while some of these approaches are costly and complex, other methods—proved to be effective—are quite straightforward, simple, and low in cost.

The seventeen papers in this volume look at a wide variety of mitigation efforts, from simple and inexpensive to considerably complex and costly. They examine techniques applied to large collections and to small house museums and to exhibition cases containing objects, as well as to single monumental works of art or historic structures. Approaches range from securing and restraining objects to decoupling them from the ground through a variety of base isolation mechanisms. All, in their own ways, are pioneering efforts in this ongoing initiative. And all of these efforts have been developed in the face of significant challenges, since—as any engineer, conservator, or mount maker who has undertaken this work will attest—a small sculpture can often be a far greater challenge to protect than a multistory building.

This publication does not address post-earthquake response or the secondary effects of earthquakes, such as fire, flooding, or landslides. While preparations against these threats and the vital responses and recovery efforts in their wake will continue to gain from further development, they are already well represented in the preservation/conservation literature. What is lacking in the literature and what this volume hopes to contribute is attention to the objects, the collections, the contents that are inside the envelope we call the museum. It is a focus much needed in the field of preservation, since to date the majority of research on seismic mitigation has been applied to structures. None, however, would deny the importance of protecting the envelope, especially since such efforts protect lives and preserve historic structures. Indeed, in his contribution, O. P. Agrawal notes that if the building collapses, the collections will surely be damaged and may not be retrievable. Further damage from exposure, and even from rushed efforts to evacuate the collections, often follows, once the envelope is ruptured. Several authors refer to the importance of building stabilization, whether it be a historic building or a purpose-built museum structure. Donna Strahan reports on the complete isolation of a museum building as an effort to protect lives, the building itself, and the collections within. Charles A. Kircher and Alex G. Tarics offer an overall review of the effectiveness of building isolation, and William S. Ginell presents several case studies on the protection of highly fragile historic adobe structures. The developments in these areas have direct implications for the protection of individual objects and collections through base isolation.

Any effort to protect cultural objects from damage begins with understanding the forces or agents that threaten harm, the effects they may have on the object, and the ways in which those effects might be mitigated or even eliminated. Mitigation starts with accepting the degree of likelihood of each of these events, then understanding their nature and the secondary effects they cause. Paul G. Somerville and Robert W. Graves, as well as Constantine Spyrakos, Ioannis Koutromanos, and Charilaos Maniatakis, pursue this understanding in their contributions; one paper takes a purely seismological approach, and the other pursues an effort to characterize the response of representative categories of object types. The benefits of applying this information and of assuring that it is regularly reviewed and updated are clearly demonstrated in the work carried out by Matthew J. Schoettler and Andreas Stavridis to verify the effectiveness and limitations of the base isolators used at the Getty Museum.

It should be kept in mind, of course, that objects are rarely damaged by the direct transmission of earthquake forces, even though such forces can be significant. Damage is more often the result of dynamic motions that are initiated. Rocking, overturning, tipping, and collisions most often impart the greatest forces and do the greatest damage. The efforts in minimizing these effects and mitigating the chain of events lead to the protection of vulnerable objects.

Providing suitable mounts for exhibited objects is one of the most important preventative actions that museums and collections in seismically active areas can undertake. A well-designed seismic mount does not necessarily have to be large, complex, or costly to be effective. This fact is well demonstrated in the contribution of Nancy Mendez, Karen Hildebrand, and Robert Sieger, who present examples of impressively effective mitigation efforts in small historic house museums. What is always necessary, as all of the authors mention many times, is that restraints and supports must be conceived, built, and attached with the necessary degree of background information about the site and the object, including the basic structural and response characteristics of the building, the gallery, and the display furniture, as well as the object's material characteristics and condition. A clear understanding of the characteristics of the materials used to anchor and restrain objects is also necessary. Anne Crowley and Debra F. Laefer provide a thorough study of the strength and appropriate use of wax, a material commonly used to anchor objects in museum displays. Theirs is a much-needed study that will contribute to the effective use of a low-cost solution.

Consideration of the basic laws of physics and a good deal of common sense are also important tools in this effort. McKenzie Lowry and colleagues provide an overview of the application of both in instances ranging from simple tie-down monofilament restraints to complex and evolving base isolation units for objects and display cases. Although more complicated and costly, base isolation can be an effective measure for protecting highly fragile and unstable objects without undue direct intervention to the object or unsightly and large support structures. The concept of base isolation for structures has been around for quite a long time. In the modern era, one of the earliest examples is the "double concave rolling ball bearing isolator," patented in 1870 by Jules Touaillon. While materials have changed and alternate designs have been introduced, the basic operating premise remains fairly consistent and has begun to be introduced in a number of collections and historic structures over the last several decades. Kimio Kawaguchi from Japan, Mustafa Erdik from Turkey, John Hirx and McKenzie Lowry and colleagues from the United States, and Vlasis Koumousis from Greece all address a variety of base isolation designs and applications. Koumousis adds to this an impressively thorough and state-of-the-art analysis of the object being isolated.

The efforts reported in this volume describe both an increasing sophistication on the part of conservation professionals when facing the challenges of protecting collections from earthquake forces and a rise in productive collaborations between conservators, mount makers, seismic engineers, and seismologists. That's good news for us all. Such collaborations are well illustrated by the impressive work of Nicholas Dorman at the Seattle Art Museum. The bad news (as the ubiquitous equation of balance reminds us) is that earthquakes are permanent phenomena on our planet. They have been with us since time immemorial. The Greek geographer and author Strabo described the effects of many earthquakes in the fifth and fourth centuries B.C. Yet these events appear relatively modern as one begins to look at the geological record. We can be certain more devastation is to come, and we must undertake efforts to protect our priceless material heritage. Returning again to the good news, it is clearly possible to prepare for and to mitigate the effects of this inevitable threat. A new generation is taking notice of the importance of these challenges and is bringing some very creative solutions to the mix; such solutions are admirably represented by the work of Victoria McGuinness.

A key aspect of success in this effort is the support of museum directors, curatorial staff, and designers. Such support is particularly vital when a museum is coping with the time constraints and pressures of special, short-term exhibitions that include loan objects. This particular challenge is described fully by Brian Considine and Ziyad Durón. Without such support, the complex challenges of reducing damage caused by earthquakes can quickly become a series of overwhelming roadblocks—making the threat that much easier to ignore than to face. The continuing interest of engineers and seismologists is also critical, as is their willingness to face unusual demands in arenas somewhat foreign to engineering standards and assumptions.

I would like to express my gratitude to Michael Brand for his support of the conference and of this publication. It takes a great deal of foresight to realize that, while the business of making artworks accessible is the most visible aspect of any museum, it is accompanied by the responsibility to assure the safety of the collections. The efforts required to face a threat that is unpredictable, less than perfectly understood, and often overwhelming to us all benefit immeasurably from such support.

Additional thanks go to the many staff members of the Getty Museum at the Villa, whose commitment to this project has been unwavering and who share in the excitement of crossing boundaries to find new solutions. We started with an equation: *acceptable risk = acceptable damage*. If this formula holds true, then it follows that as we reduce the risk, we will reduce the damage.

New Developments in Seismology: The Ground Motion Environment of the Los Angeles Region

Paul G. Somerville Robert W. Graves

Introduction

The earthquake environment of the Los Angeles region is highly complex. In addition to the San Andreas Fault and the system of strike-slip faults that are aligned roughly parallel with it, there is a system of thrust faults that have produced the Transverse Ranges. These thrust faults include both faults that intersect the ground surface and blind-thrust faults that do not. Examples of recent earthquakes that have occurred on blind-thrust faults that do not break the ground surface include the 1983 Coalinga, 1987 Whittier Narrows, and 1994 Northridge earthquakes. Disaggregation of the probabilistic seismic hazard for the Los Angeles region in the United States Geological Survey (USGS) National Seismic Hazard Maps shows that, at the annual probabilities that apply to building codes, the hazard is dominated by earthquakes having moment magnitudes (M_w) of about 7.0 and closest distances of about 10 km (6.2 mi.). At periods of several seconds, a significant contribution also comes from larger, more distant earthquakes on the San Andreas Fault.

Blind-thrust faults were not included in the fault maps developed in 1998 by the California Division of Maps and Geology (CDMG), but if they had been, then there would be very few locations in the Los Angeles area located more than 10 km from the surface trace or the vertical surface projection of an active fault. Thus, the ground motion environment in the Los Angeles region is dominated by the near-fault conditions of moderate-magnitude earthquakes.

Spatial Variability of Near-Fault Ground Motions

Experience gained in the course of deriving earthquake rupture models from strong motion recordings of many earthquakes has shown that the large spatial variations that are observed in the ground motions recorded around faults, especially those at periods of 1 sec. and longer, can usually be attributed to features of the earthquake source such as the orientation of the fault plane, the style of faulting (strike slip or dip slip), and the evolution and distribution of slip on the fault plane. For example, the rupture directivity effect (the generation of a strong pulse of ground motion that is observed at stations toward which rupture propagates but not at other stations) can be readily explained with seismological models. These models are capable of reproducing the waveforms of recorded ground motions having periods of 1 sec. and longer, which are important for tall buildings.

The first attempts to represent the spatial variation of ground motions around faults due to earthquake source effects in ground motion attenuation models have been made in the last decade. The motivation for incorporating these effects in ground motion models is to increase the accuracy and reduce the variability associated with the median ground motion predicted by the model. This variability in the ground motion has a large effect on ground motion estimates used for seismic design. This finding is true whether the design ground motions are based on a deterministic approach that uses the median plus one standard deviation level, or on a probabilistic approach that incorporates the variability in the ground motion. The first models representing predictable spatial variation of ground motions around faults include a model of rupture directivity effects (Somerville et al. 1997) and a model of the differences in ground motions between the foot wall and hanging wall of dip-slip faults (Abrahamson and Somerville 1996; Abrahamson and Silva 1997). These preliminary models were derived empirically from recorded strong motion data using conceptual models of these phenomena to guide the form of the ground motion models.

To date, these models have not been systematically applied in the generation of the USGS National Seismic Hazard Maps. Consequently, the USGS maps, upon which the *International Building Code* (*IBC*) maximum considered earthquake (MCE) design-ground-motion maps are based, do not include the spatial variability in ground shaking level that is expected from these effects. Moreover, the approach of truncating the probabilistic ground motion for an annual probability of occurrence of 1/2500 on site B (a soft rock site condition) at levels of 1.5 g and 0.6 g for periods of 0.2 and 1.0 sec., respectively, unless it is exceeded by 1.5 times the median ground motion level for the maximum earthquake, has the effect of smoothing out the spatial variations caused by these effects.

Near-Fault Rupture Directivity Pulse

Near-fault ground motions are different from ordinary ground motions in that they often contain strong coherent dynamic long-period pulses and permanent ground displacements. The dynamic motions are dominated by a large long-period pulse of motion that occurs on the horizontal component perpendicular to the strike of the fault, caused by rupture directivity effects (Somerville et al. 1997). Near-fault recordings from recent earthquakes indicate that this pulse is a narrow-band pulse with a period that increases with magnitude (Somerville 2003). In figures 1a and 1b, rupture directivity pulses of earthquakes in the magnitude range of M_w 6.7 to 7.0 are compared with pulses from earthquakes in the magnitude range of 7.2 to 7.6. The narrow-band nature of these pulses causes their elastic response spectra to have peaks. The fault-normal components of the response spectra (which contain the directivity pulse) are shown as solid red lines, and the fault-parallel components (which, as expected, are much smaller at long periods) are shown by long dashed blue lines. The 1994 *Uniform Building Code* (*UBC*) spectrum for soil site conditions is used as a reference model for comparison (International Conference of Building Officials 1997). The fault-normal spectra for the large earthquakes (right column) are compatible with the *UBC* code spectrum in the intermediate period range, between 0.5 and

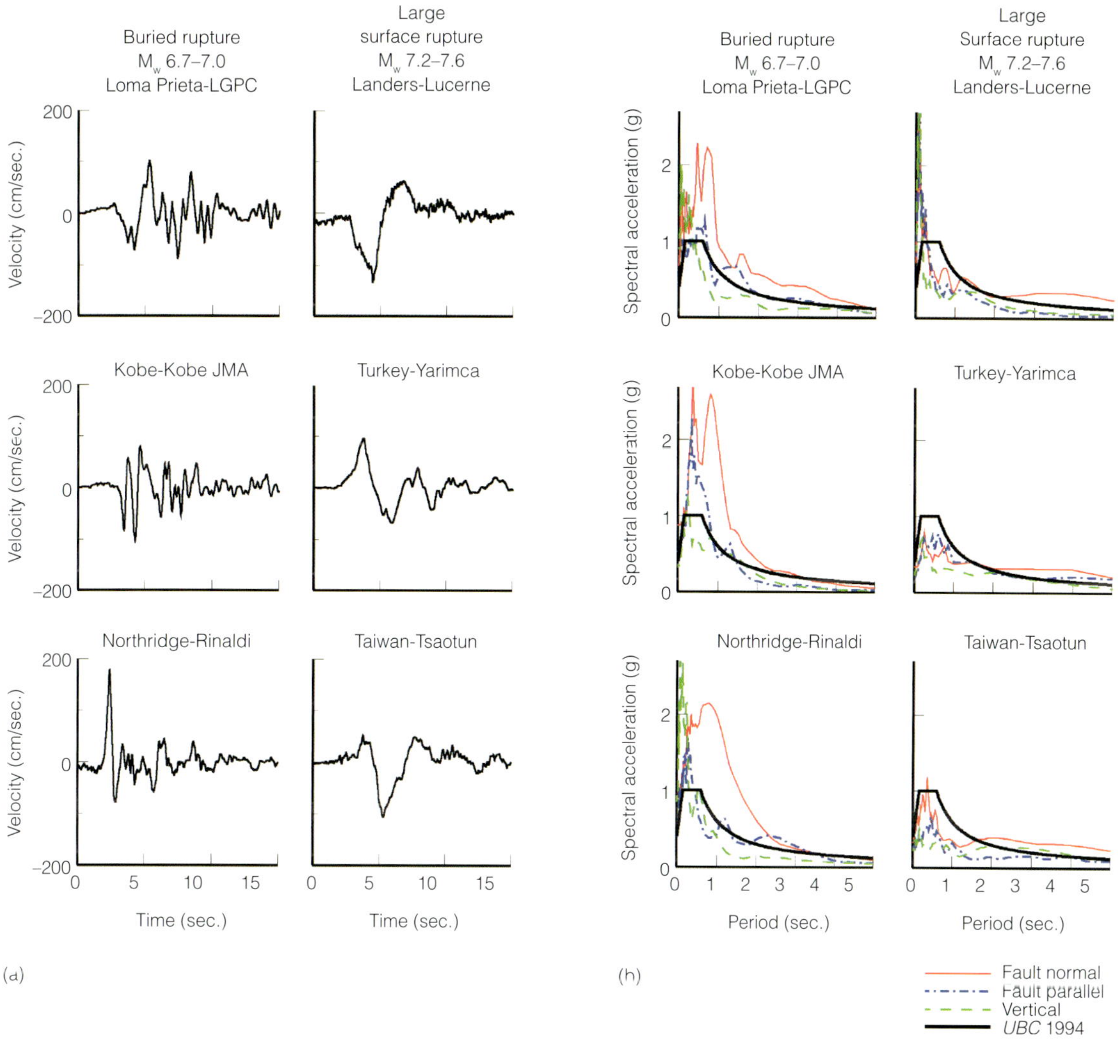

2.5 sec., but have a bump at a period of about 4 sec., where they significantly exceed the 1994 *UBC* code spectrum. The fault-normal spectra of the smaller earthquakes (left column) are very different from those of the larger earthquakes. Their spectra are much larger than the 1994 *UBC* code spectrum in the period range of 0.5–2.5 sec., but they are similar to the *UBC* spectrum at longer periods. This magnitude dependence of the pulse period causes the response spectrum to have a peak with a period that increases with magnitude, such that the near-fault ground motions from moderate-magnitude earthquakes may exceed those of larger earthquakes at periods of 1 to 2 sec. (Somerville 2003). The rupture directivity model of Somerville et al. (1997) contains a broadband adjustment for rupture directivity effects, which generally envelops the narrow-band model of Somerville (2003), except at short periods (about 1 sec.).

Figures 1a and 1b
Fault-normal velocity pulses recorded near three moderate-magnitude earthquakes (left column) and three large earthquakes (right column), shown on the same scales (a). Corresponding acceleration response spectra, with the 1994 *UBC* code spectrum for soil conditions shown for comparison (b). (LGPC = Los Gatos Presentation Center, a strong motion recording station, JMA = Japan Meteorological Agency.) Source: Somerville 2003.

The 1997 *UBC* incorporated a near-source factor that provides, in an approximate way, for the rupture directivity effect and the hanging wall effect. However, the exclusion of the blind-thrust faults from the near-source maps (CDMG 1998) results in near-fault effects being treated incompletely.

Reverse-Faulting Earthquakes and Hanging Wall Effects

The ground motions from reverse-faulting earthquakes, such as the 1994 Northridge Earthquake, are systematically stronger on average, by a few tens of percent, than the ground motions from strike-slip earthquakes. This effect is incorporated into the ground motion attenuation models that are in current use, and in the USGS National Seismic Hazard Maps. However, the ground motions on the hanging wall of these faults are systematically larger than at other locations, especially at periods shorter than 1 sec., but this is not taken into account in most ground motion attenuation relations.

Buried-Faulting Earthquakes

The ground motions from earthquakes that do not break the ground surface are systematically stronger than the ground motions that do break the surface (Somerville 2003; Somerville and Pitarka 2006). This finding was demonstrated in the 1994 Northridge Earthquake, which occurred on a blind-thrust fault with a top edge located at a depth of about 5 km (3.1 mi.); the event produced ground motions that are about 20% stronger than those of the average reverse-faulting earthquake. The same conditions appear to apply to strike-slip and oblique-faulting earthquakes. The 1989 Loma Prieta Earthquake, which occurred on an oblique fault rupture with a top edge located at a depth of 3.5 km (2.2 mi.), produced ground motions about 30% stronger than the average strike-slip earthquake. The large differences in ground motion levels between surface and buried faulting events are evident in figures 2a and 2b, which show the response spectra of near-fault recordings of recent large earthquakes. The left panel shows recordings from four surface-faulting earthquakes in the M_w range of 7.4 to 7.9, and the right panel shows recordings from two buried-faulting earthquakes of magnitudes M_w 6.7 and 7.0. The response spectra of the deep earthquakes are much stronger than those of the larger shallow earthquakes for periods less than 1.5 sec.

Figure 3 shows the event terms for larger sets of surface-rupture earthquakes at the top; subsurface-rupture earthquakes are shown at the bot-

Figures 2a and 2b
Near-fault response spectra of recent large earthquakes. Four earthquakes, M_w 7.4 to 7.9, with shallow asperities (regions on a fault with larger-than-average fault slip) and large surface faulting (a). Two earthquakes, M_w 6.7 and 7.0, with deep asperities and no surface faulting (b). Source: Somerville 2003.

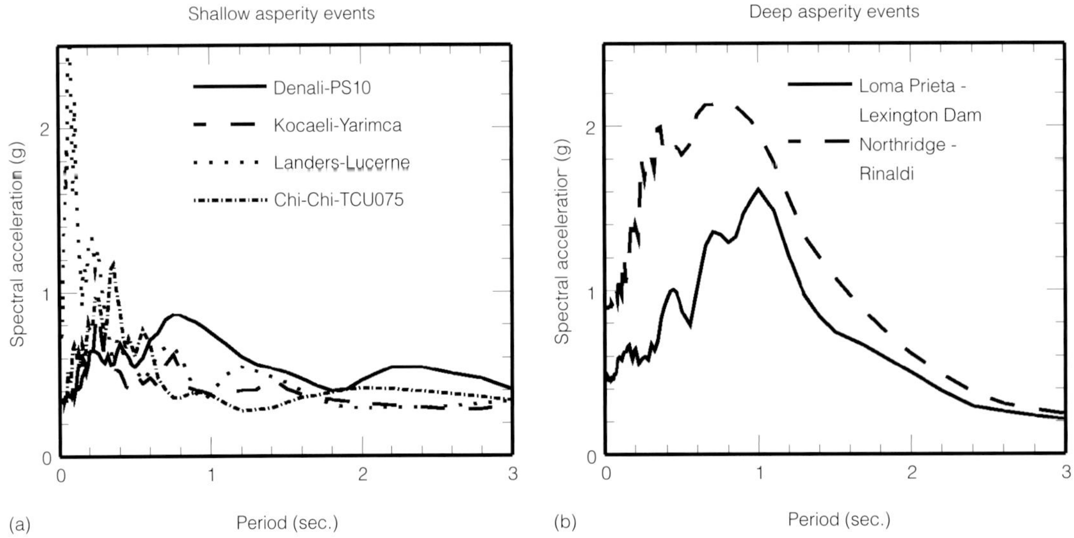

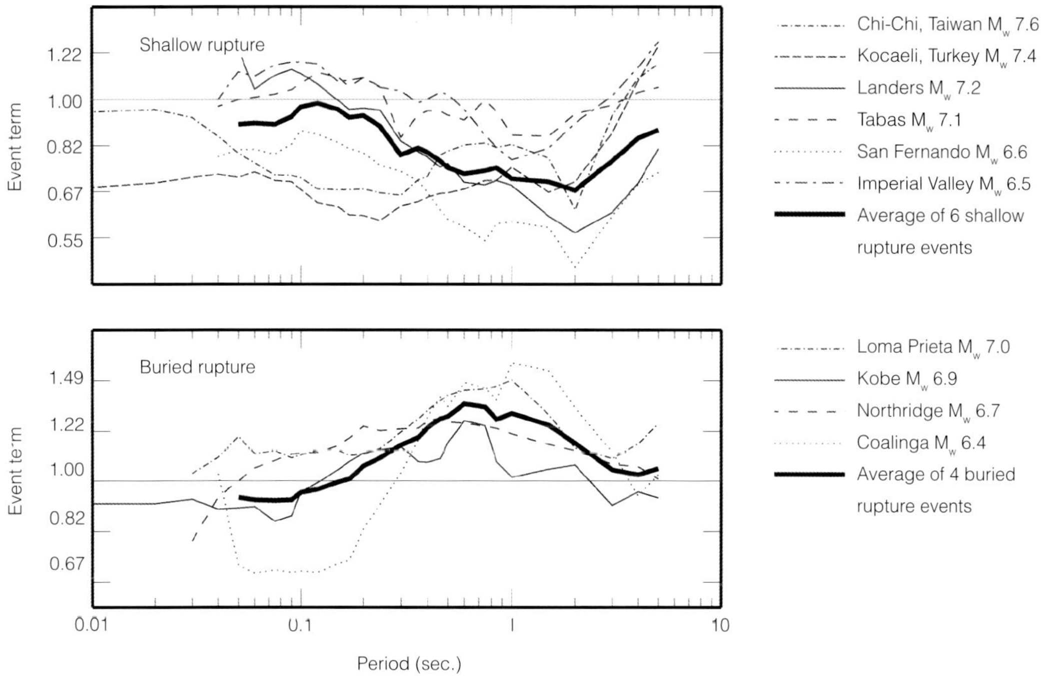

Figure 3
Comparison of response spectral amplitude of individual earthquakes with surface rupture (top) and buried rupture (bottom), averaged over recording sites, with the amplitude of the average earthquake as defined by the model of Abrahamson and Silva (1997), represented by the unit line, which accounts for magnitude, closest distance, and recording site category. The event terms (residuals) are shown as the ratio of the event to the model. Source: Somerville 2003.

tom. The unit line represents the Abrahamson and Silva (1997) model, and lines above this line indicate that the event's ground motions, on average, exceed the model (Abrahamson, Somerville, and Cornell 1990). The event terms of the 1999 Turkey and Taiwan earthquakes are represented by their residuals from the model. The ground motions of the subsurface-rupture earthquakes are systematically stronger than average, and those of the surface-rupture earthquakes are weaker than average, over a broad period range centered at 1 sec., which dominates peak velocity. At periods between 0.3 and 3.0 sec., the ground motions from earthquakes that produce large surface rupture are only about one-half as strong as those in which rupture is confined to the subsurface.

Ground Motions from Large Surface-Faulting Earthquakes

Recent large earthquakes that had large surface slip, including the Chi-Chi, Taiwan; Denali, Alaska; and Kocaeli, Turkey, events, have surprisingly weak ground motions at short and intermediate periods (Somerville 2003). At periods between 0.3 and 3.0 sec., the ground motions from earthquakes that produce large surface rupture are only about one-half as strong as those in which rupture is confined to the subsurface, and are several tens of percent weaker than predicted by current ground motion models, as shown in the upper part of figure 3. At longer periods, the ground motions become comparable to those of ground motion models in current use. A new set of ground motion attenuation relations is being developed by the Pacific Earthquake Engineering Research Center (PEER) Lifelines Next Generation Attenuation (NGA) Project. Incorporation of the strong motion recordings of these recent large earthquakes into the NGA ground motion attenuation relations results in partial magnitude saturation, in which the ground motion level almost stops increasing with magnitude for magnitudes larger than about M_w 7.0 for periods less than about 3 sec. These changes will have a much larger impact in the San Francisco Bay Area, where large surface-faulting earthquakes on the San Andreas and Hayward faults dominate the seismic hazard, than in the Los Angeles region, where the hazard is dominated by smaller earthquakes, some of which occur on blind thrusts.

Basin Effects on Ground Motions

Many urban regions are situated on deep sediment-filled basins. A basin consists of alluvial deposits and sedimentary rocks that are geologically younger and have lower seismic wave velocities than the underlying rocks upon which they have been deposited. Basins have thickness ranging from 100 m (328 ft.) to over 10 km (6.2 mi.). It is widely recognized that sedimentary basins have a strong influence on strong ground motions, especially at periods longer than about 1 sec. However, most empirical ground motion models do not explicitly account for these effects.

Current building code criteria for characterizing site response, including the 1997 *UBC* (International Conference of Building Officials 1997) and the 2003 *NEHRP Recommended Provisions for Seismic Regulations for New Buildings and Other Structures* (Building Seismic Safety Council 2004), are based on the average shear wave velocity within the upper 30 m (98.4 ft.). However, at periods longer than 1 sec., seismic waves have wavelengths much longer than 30 m, with amplitudes controlled by geological structures having depths of hundreds or thousands of meters, so we expect that seismic velocity structure that is much deeper than 30 m will influence the amplitudes of long-period seismic waves. In most cases, such as in sedimentary basins, this deeper geological structure is not horizontally layered. If the wave is propagating in the direction in which the basin is thickening and enters the basin through its edge, it can become trapped within the basin if post-critical incidence angles develop.

A model that predicts basin amplification effects was developed through a collaborative program of studies sponsored by the Southern California Earthquake Center (SCEC) and the PEER Lifelines project. Five groups of researchers participated in extensive testing of procedures for simulating ground motions in basins using 3-D finite difference and 3-D finite element methods. This process resulted in the virtual elimination of discrepancies among the five simulation procedures. These groups then participated in the simulation of six different earthquake rupture scenarios on each of ten faults in the Los Angeles region (Day et al. 2006). The results of these simulations were used to develop relationships between ground motion amplification and the depth of the basin, expressed as the depth to 1.5 km/sec. (0.93 m/sec.) for response spectral amplitudes in the period range of 2–10 sec. These amplification factors are shown in figure 4, which uses a 500 m (1640 ft.) depth related to a shear wave velocity of the medium (V_s) of 1.5 km/sec. as a reference to represent empirical ground motion models for rock.

Basin Edge Effects

Large amplification of ground motions occurs when seismic waves enter basins having steep fault-controlled margins. For example, in the 1994 Northridge Earthquake, the abrupt deepening of the Los Angeles basin to a

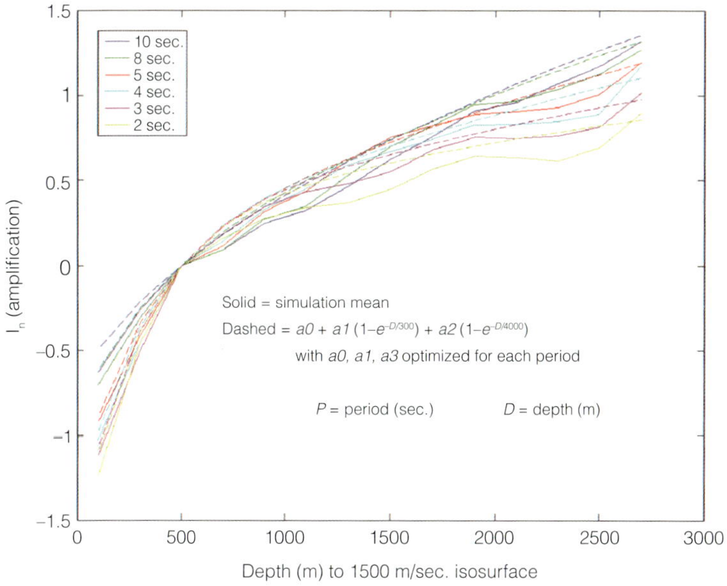

Figure 4
Amplification (I_n, or natural logarithm) as a function of basin depth (depth to V_s 1.5 km/sec., or 0.93 mi./sec.) for ground motion periods ranging from 2 to 10 sec., based on simulations of six different earthquake scenarios on each of ten faults in the Los Angeles region. A depth of 500 m (1640 ft.) is used as a rock site reference for empirical ground motion models. Source: Day et al. 2006.

depth of several kilometers that occurs at the Santa Monica Fault (fig. 5, left side) caused the constructive interference of two arrivals: the waves entering the basin from the side, and body waves entering the basin from below, which combined to produce the basin edge wave (Graves, Pitarka, and Somerville 1998). The time histories recorded in the Los Angeles basin have much larger amplitudes and longer durations than those recorded on rock sites in the Santa Monica Mountains (fig. 5, right side). These large waves consist of surface waves that have become trapped in the Los Angeles basin. South of the Santa Monica Fault, even greater amplification occurs where the basin suddenly deepens across the fault. The large long-period ground motions recorded at Santa Monica are due to the basin edge effect. The basin edge effect is confined to a zone that is on the order of a few kilometers wide, lying south of the Santa Monica Fault and marked by the red-tagged buildings.

Much of the damage in the Los Angeles basin during the Northridge Earthquake, including the collapse of the I-10 freeway and damage to numerous large buildings in Santa Monica and West Los Angeles, is attributable to basin effects and basin edge effects. The basin edge effect is caused by an abrupt lateral contrast in shear wave velocity caused by faulting. Basin edges that have smooth, concave bedrock profiles not controlled by faulting are not expected to cause basin edge effects. Thus, we expect that basin edge effects are not a general feature of basins but are, instead, confined to particular kinds of basin edges.

Puente Hills Blind-Thrust Earthquake Scenario

We can incorporate all of the ground motion effects described above by using seismological procedures for the simulation of strong ground motion. We demonstrate these procedures using earthquake scenarios on the Puente Hills Blind Thrust. Recent studies have identified the seismic potential of the Puente Hills fault system, which lies directly beneath downtown Los Angeles. The geometry and slip rate of the system were defined by Shaw et al. (2002), and four large-slip paleoearthquakes on this system were identified by Dolan,

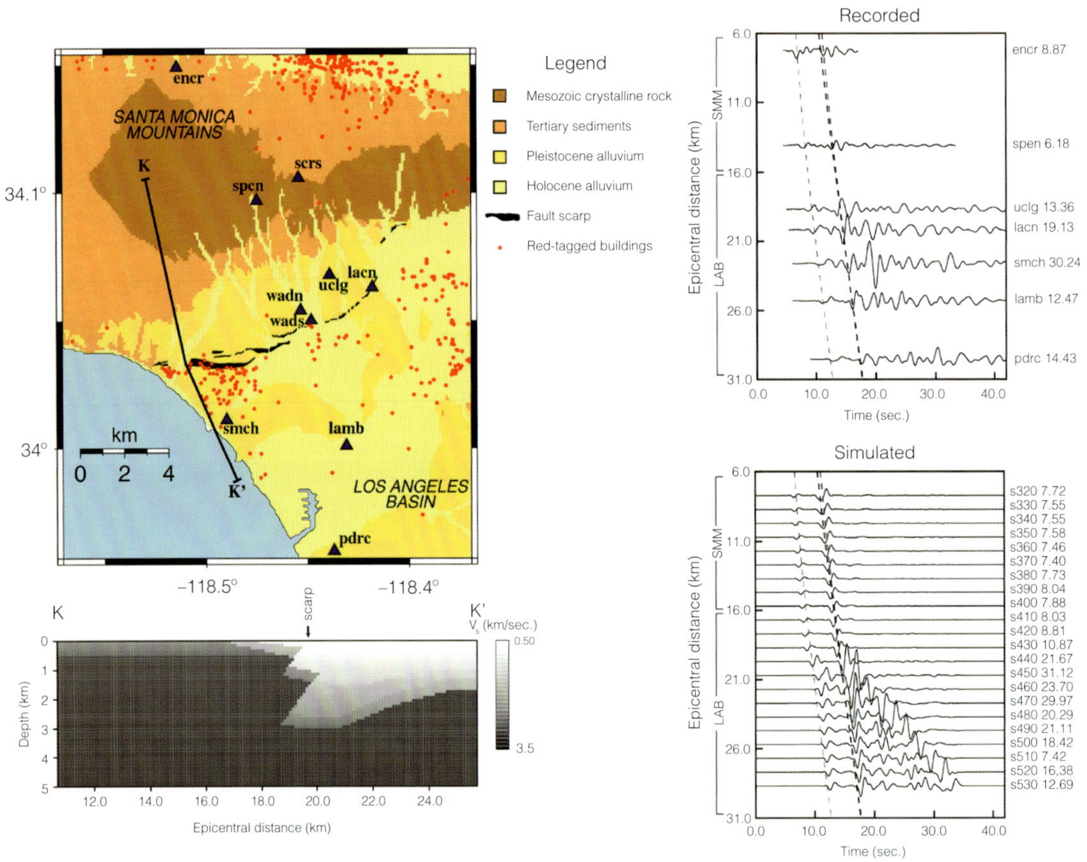

Figure 5
Surface waves generated in the west Los Angeles basin by the 1994 Northridge Earthquake. Source: Graves, Pitarka, and Somerville 1998.

Christofferson, and Shaw (2003). The seismic potential of this fault system was manifested by the 1987 Whittier Narrows Earthquake, which occurred on the Santa Fe Springs segment of the Puente Hills Blind Thrust (Shaw et al. 2002). Figure 6 shows a map of the Los Angeles region indicating the location of the Puente Hills fault system.

Particular ground motion issues that need to be addressed include the effects of forward rupture directivity, which will be strongest in regions, including downtown Los Angeles, that lie above the top edge of the Puente Hills Fault; the effects of basin-trapped waves, which will be strongest in the Los Angeles basin to the south of the fault, including Long Beach; and hanging wall effects, which will be strongest in the northern part of the Los Angeles basin (including downtown Los Angeles). Current empirical ground motion models, such as those used by Field et al. (2005), do not provide realistic representations of all of these effects. More realistic representations of the ground motions that could be generated by earthquakes occurring on the Puente Hills Blind Thrust can be obtained through the use of broadband strong motion simulation procedures.

The authors used full waveform modeling to further quantify the ground motions that could be generated by various scenario earthquakes occurring on the Puente Hills system and assess their impacts on buildings (Naeim and Graves 2006). To do this, we calculated broadband ground motion time histories for the prescribed scenarios using a hybrid simulation technique (Graves and Somerville 2006). The procedure used to simulate broadband

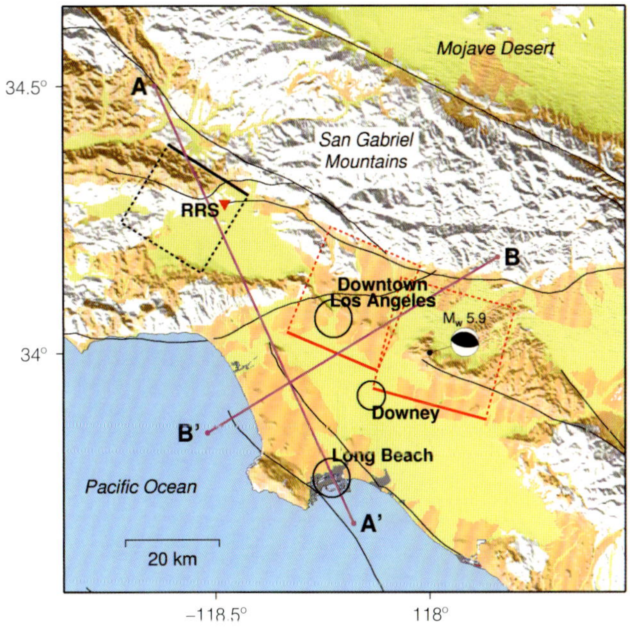

Figure 6
Map of the Los Angeles region indicating the surface projection of the Northridge fault plane (black) and the Puente Hills fault system (red), as well as the location of the Whittier Narrows Earthquake (M_w 5.9). The Los Angeles segment of the Puente Hills system lies directly beneath downtown Los Angeles. Source: Graves and Somerville 2006.

ground motions of Puente Hills Blind Thrust earthquakes is a hybrid procedure that combines a stochastic approach at high frequencies and a deterministic approach at low frequencies. A detailed description of this methodology is given in Graves and Pitarka (2004).

Three rupture scenarios for the Puente Hills system were considered. The first scenario (R-1) is an M_w 6.7 event that ruptures only the Los Angeles segment of the system. The second scenario (R-2) is an M_w 7.2 event that ruptures the entire system (represented as two segments). The third scenario (R-3) uses the same fault segments and slip distribution as R-2; however, the rise time is doubled, and the rupture velocity is reduced to 75% of the local V_s. Scenario R-3 is designed to represent a low dynamic stress-drop event. We calculate broadband (0–10 Hz) ground motions over an area 110 by 150 km (68.2 × 93 mi.), which spans most of the greater Los Angeles metropolitan region. This region is divided into a uniform grid of sites with a spacing of 500 m (1640 ft.), yielding a total of 66,000 sites. The low-frequency ($f < 1$ Hz) motions are calculated on a fine mesh ($h = 125$ m, or 410 ft.) using a finite-difference (FD) approach, and then subsampled to the coarse grid. Nearly four hundred million grid points are needed for the FD model, and the computations were performed on the Linux High Performance Computing Cluster at the University of Southern California. The high-frequency ($f > 1$ Hz) motions are calculated by use of the stochastic approach applied individually for each of the 66,000 locations.

Figure 7 displays snapshots of the propagating broadband velocity wave field for scenario R-2. The complexity of the propagating wave field is readily apparent in these plots. Strong rupture directivity channels large-amplitude pulses of motion directly into the Los Angeles basin, which then propagate southward as basin surface waves. Due to the southward rupture, virtually all of the strongest shaking is directed south from the fault.

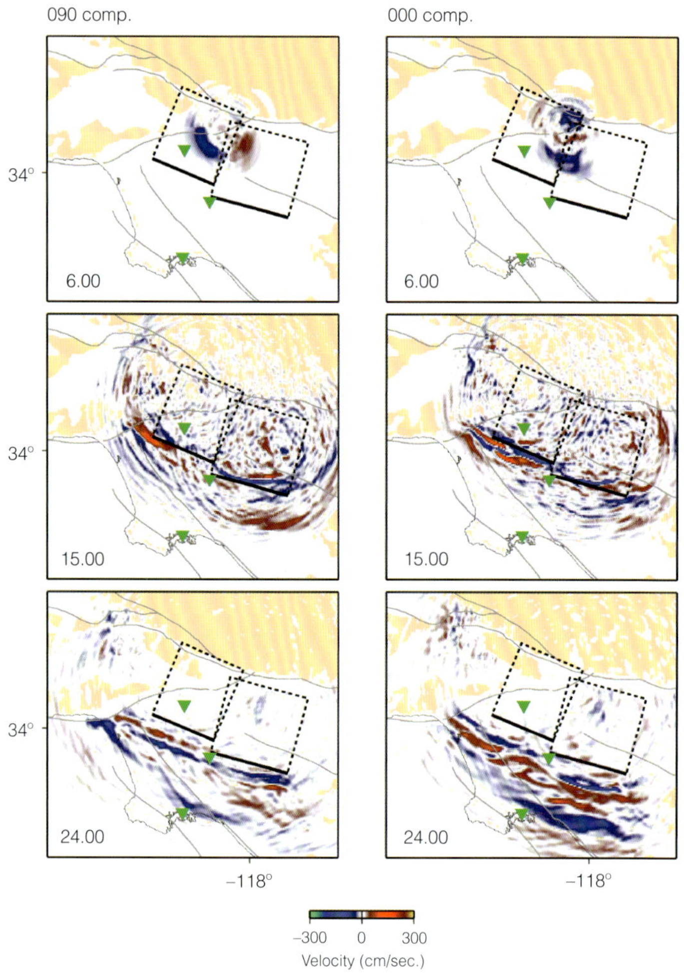

Figure 7
Snapshots of the broadband (0–10 Hz) ground velocity for the E-W (left) and N-S (right) components of motion for scenario R-2. Time (in sec.) after earthquake initiation is indicated in the lower left of each panel. The black rectangles indicate the surface projection of the fault planes. Tan shading indicates surface exposure of more consolidated rock; white regions are primarily sediment-filled basins. The green triangles denote locations (top to bottom) for downtown Los Angeles, Downey, and Long Beach. Source: Graves and Somerville 2006.

This complexity is further apparent in comparing the broadband ground velocity time histories simulated at downtown Los Angeles, Downey, and Long Beach for all three rupture scenarios (fig. 8). In all cases, the waveform at Los Angeles is dominated by rupture directivity, which produces a strong, concentrated pulse of motion. At Downey, complexity begins to develop in the waveforms with the appearance of secondary pulses of motion. These pulses are surface waves that are beginning to develop in the basin structure. At Long Beach, which is located across the Los Angeles basin from the rupture, the waveforms are dominated by late-arriving longer-period surface waves. These surface waves are generated in the near-fault region, as the direct waves from the rupture are trapped within the sediments of the basin. In general, the characteristics of the motions at each site do not vary greatly for these three scenarios. The most significant effect is the relatively lower amplitude motions generated by scenario R-3.

The high density of sites used in the calculations allows us to construct detailed maps of various ground motion parameters. Examples of these types of ground motion maps are shown in figure 9, which displays peak ground acceleration (PGA) and spectral acceleration (SA) at 1 sec. and at 3 sec. for each of the three scenarios. At high-frequency PGA, the pattern of ground motions

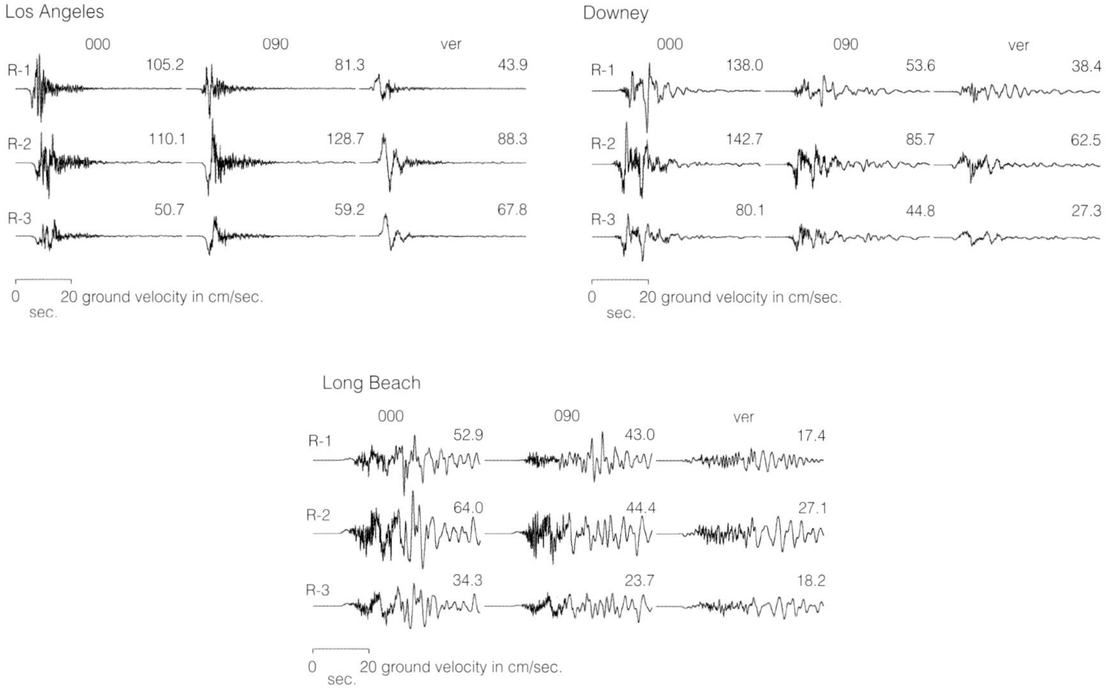

Figure 8
Broadband ground velocity time histories simulated at downtown Los Angeles, Downey, and Long Beach for the three rupture scenarios (000 = north, 090 = east, and ver = vertical component). Source: Graves and Somerville 2006.

is centered over the fault rupture planes, with the largest motions typically occurring on the hanging wall. At longer periods (1 sec. and 3 sec. SA), the strongest motions occur along the top edge of the rupture planes and are the result of rupture directivity effects. These strong pulses of motion are channeled directly into the soft sediments of the Los Angeles basin and become trapped as basin-generated surface waves. These large-amplitude surface waves travel southward across the basin, giving rise to an elongated pattern of elevated, longer-period response that extends across the basin.

Scenario R-2 produces the largest overall ground motion levels, followed by scenario R-1 and then by R-3. Both scenarios R-1 and R-2 have been modeled as buried rupturing events with relatively short slip-rise times (Somerville et al. 1999). The use of a lower rupture velocity and longer rise time in scenario R-3 (i.e., relatively low dynamic stress drop) significantly reduces the ground motion levels compared to scenario R-2, even though the slip distribution and total seismic moment are the same.

Both scenarios R-2 and R-3 produce larger motions in the eastern portion of the Los Angeles basin than in the western portion of the basin. This difference is due primarily to the prescribed slip distribution, which has the strongest asperity (larger-than-average fault slip) on the eastern fault segment. We expect that this ground motion pattern would be reversed if the slip distribution for the two segments were switched. For all three scenarios, the combination of rupture directivity and basin response produces elevated levels of long-period ground motions throughout the Los Angeles basin region. Although the high-frequency motions are less sensitive to these effects, the PGA levels predicted in our simulations are still large, exceeding 1 g in the near-source region over the fault planes in all cases. For these scenarios, the most significant variations are seen in the reduced ground motion levels predicted by the low

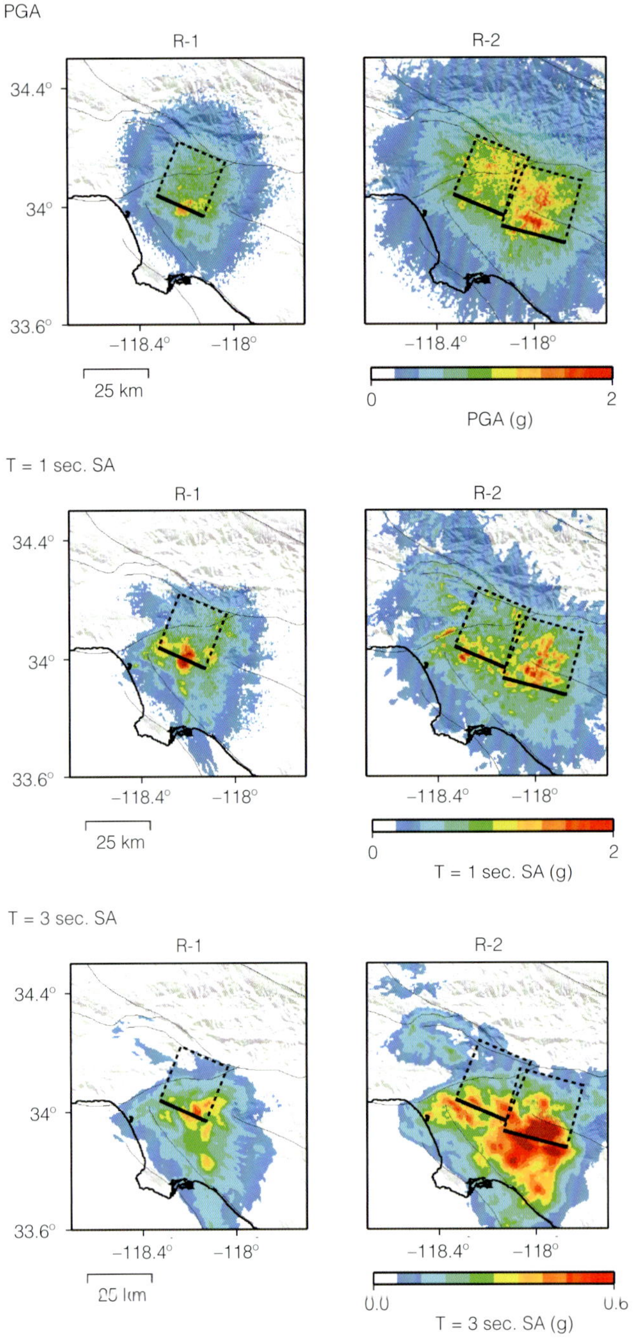

Figure 9
Ground motion maps for peak ground acceleration (PGA) and response spectral acceleration (SA) at 1 sec. and at 3 sec. for the three rupture scenarios (R-1, R-2, and R-3). The maps in each row are all scaled to the same level to allow direct comparison among the scenarios. Source: Graves and Somerville 2006.

dynamic stress-drop event (R-3) compared to both R-1 and R-2. This finding stresses the need to consider various source parameterizations, including constraints provided by dynamic rupture analyses in the assessment of seismic hazard. Obviously, one important aspect of source characterization that we have not fully considered here is the effect of slip distribution. We expect the pattern of ground motions will be dependent on the location of asperities on the fault planes, and further work is needed to more fully quantify and understand the impact of this effect on hazard characterization.

Time-History Modeling of Probabilistic Hazard with CyberShake

In the preceding section, we described the simulation of a scenario earthquake on the Puente Hills Blind Thrust. However, in most locations in Los Angeles, the maximum considered earthquake (MCE) specified by the *IBC* code is derived from a probabilistic seismic hazard analysis (PSHA) based on an annual probability of exceedance of 1/2500, which contains contributions from earthquakes of all magnitudes occurring on all of the faults with ground motions that can impact the Los Angles area. The Puente Hills Blind Thrust is just one of those faults.

Conventional PSHA uses ground motion attenuation equations to estimate the response spectrum for each considered earthquake based on the specified magnitude, distance, and site conditions. Deterministic source and wave propagation effects such as rupture directivity and basin response can have a significant impact on near-fault ground motion levels, particularly at shaking periods longer than about 1 sec. In most cases, these effects are not explicitly included in empirical ground motion models (other than through increased uncertainty), or they are included with averaged response terms, thus potentially diminishing their effectiveness in applications such as PSHA.

Using CyberShake, which is a part of the Southern California Earthquake Center Community Modeling Environment (SCEC/CME) Project, Graves and colleagues have developed hazard curves using 3-D ground motion simulations rather than attenuation relationships (Graves et al. 2006). To calculate a waveform-based probabilistic hazard curve for a site of interest, we begin with the SCEC open-source computer platform for seismic hazard analysis (OpenSHA) implementation of the 2002 USGS National Seismic Hazard Maps (NSHMP-2002) Earthquake Rupture Forecast and identify all ruptures (excluding background seismicity) within 200 km (124 mi.) of the site of interest. We convert the NSHMP-2002 rupture definition into multiple rupture variations with differing hypocenter locations, slip distributions, and slip-time functions. Strain Green Tensors, which describe seismic wave propagation, are calculated for the site using the SCEC Community Velocity Model 3.0. Then, using reciprocity, we calculate synthetic seismograms for each rupture variation. The resulting suite of synthetics is processed to extract peak intensity measures of interest. The peak intensity measures are combined with the original rupture probabilities to produce probabilistic seismic hazard curves for the site.

We have produced hazard curves for spectral acceleration at 3.0 sec. at about ten sites in the Los Angeles region. At low ground motion (high probability) levels, there is little difference between the CyberShake curves and attenuation relationship-based curves. For probabilities of more practical concern (e.g., 2% in 50 years), the methodologies produce similar results at some sites, such as downtown Los Angeles, shown in the top part of figure 10, but produce different results at others, such as Whittier Narrows, shown in the bottom part of figure 10. In particular, the disaggregation of the CyberShake hazard at the bottom right of figure 10 shows that a large contribution to the hazard at Whittier comes from large earthquakes on the southern San Andreas Fault, with magnitudes in the range of M_w 7.0–8.0 and distances in the range of 50–60 km (31–37 mi.). This finding is consistent with the results obtained

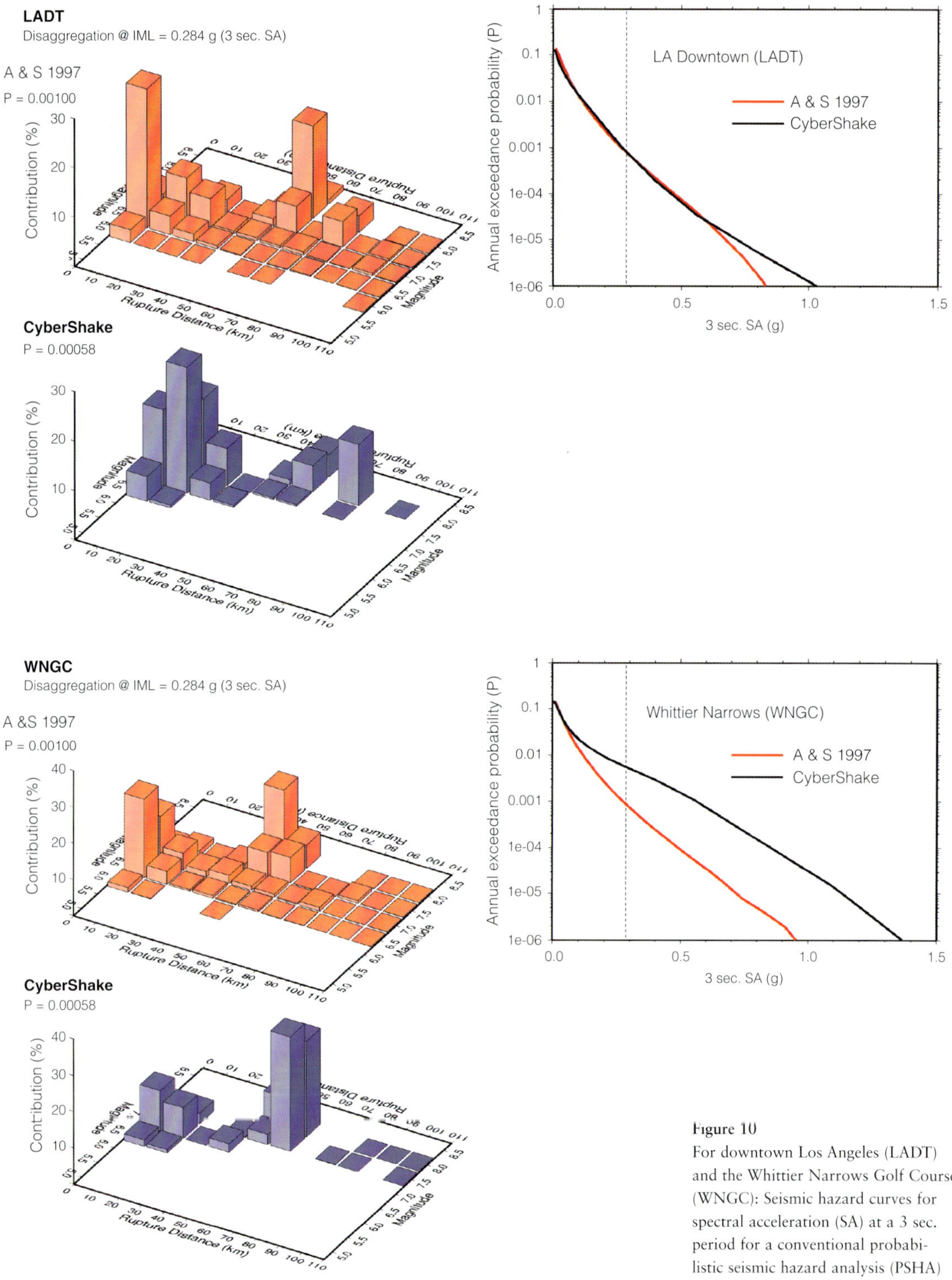

Figure 10
For downtown Los Angeles (LADT) and the Whittier Narrows Golf Course (WNGC): Seismic hazard curves for spectral acceleration (SA) at a 3 sec. period for a conventional probabilistic seismic hazard analysis (PSHA) ground motion model (lower curve) and CyberShake (upper curve) are shown on the right (Abrahamson and Silva 1997). Disaggregation of the probabilistic hazard for a ground shaking level of 0.284 g for the conventional (top) and CyberShake (bottom) PSHA calculations is shown on the left. Source: Graves et al. 2006.

for the TeraShake simulation of scenario earthquakes on the southern San Andreas Fault (Olsen et al. 2006).

Summary

We have reviewed aspects of fault geometry and geological structure that cause spatial variations in the level of ground motions expected from large-scenario earthquakes. These conditions include style of faulting, differences between surface faulting and buried faulting earthquakes, rupture directivity effects, basin effects, and basin edge effects. These effects are illustrated in large-scale ground motion simulations of scenario earthquakes on the Puente Hills Blind Thrust. The impact of these effects on probabilistic seismic hazard is also explored by means of CyberShake calculations in which the ground motions at a given site from a very large number of earthquake scenarios are represented by simulated time histories in place of conventional ground motion attenuation equations used in probabilistic seismic hazard analysis. We find that the seismic hazard estimates are quite similar at some locations, such as downtown Los Angeles, but quite different at others, such as Whittier. These differences are attributable to the coupling of rupture directivity and basin effects that are present in the CyberShake calculations but not in the conventional attenuation relations.

Acknowledgments

This work has been supported in part by the Southern California Earthquake Center (SCEC) and the external research program of the U.S. Geological Survey National Earthquake Hazards Reduction Program (NEHRP).

References

Abrahamson, N. A., and W. J. Silva
1997 "Empirical Response Spectral Attenuation Relations for Shallow Crustal Earthquakes," *Seismological Research Letters* 68, pp. 94–127.

Abrahamson, N. A., and P. G. Somerville
1996 "Effects of the Hanging Wall and Footwall on Ground Motions Recorded during the Northridge Earthquake," *Bulletin of the Seismological Society of America* 86, pp. S93–S99.

Abrahamson, N. A., P. G. Somerville, and C. Allin Cornell
1990 "Uncertainty in Numerical Strong Motion Predictions," *Proceedings of the Fourth U.S. National Conference on Earthquake Engineering* (Oakland, Calif.), pp. 407–16.

Building Seismic Safety Council
2004 *2003 NEHRP [National Earthquake Hazards Reduction Program] Recommended Provisions for Seismic Regulations for New Buildings and Other Structures,* FEMA 450-01 (Washington, D.C.).

California Division of Mines and Geology [CDMG]
1998 *Maps of Known Active Near-Source Zones in California and Adjacent Portions of Nevada* (Whittier, Calif.).

Day, S. M., et al.
2006 "Numerical Simulation of Basin Effects on Long-Period Ground Motion," *Proceedings of the Eighth National Conference on Earthquake Engineering* (Oakland, Calif.).

Dolan, J. F., S. A. Christofferson, and J. M. Shaw
2003 "Recognition of Paleoearthquakes on the Puente Hills Blind Thrust Fault, California," *Science* 300, pp. 115–18.

Field, E. H., et al.
2005 "Loss Estimates for a Puente Hills Blind Thrust Earthquake in Los Angeles, California," *Earthquake Spectra* 21, pp. 329–38.

Graves, R. W., et al.
2006 "SCEC CyberShake Platform: Incorporating Deterministic 3D Waveform Modeling into Probabilistic Seismic Hazard Curves," *Proceedings of the Seismological Society of America Centennial Meeting* (El Cerrito, Calif.).

Graves, R. W., and A. Pitarka
2004 "Broadband Time History Simulation Using a Hybrid Approach," *Proceedings of the Thirteenth World Conference on Earthquake Engineering* (Vancouver).

Graves, R. W., A. Pitarka, and P. G. Somerville
1998 "Ground Motion Amplification in the Santa Monica Area: Effects of Shallow Basin Edge Structure," *Bulletin of the Seismological Society of America* 88, pp. 1224–42.

Graves, R. W., and P. G. Somerville
2006 "Broadband Simulations for Scenario Earthquakes on the Puente Hills Fault," *Proceedings of the Eighth National Conference on Earthquake Engineering* (Oakland, Calif.).

International Conference of Building Officials
1997 *Uniform Building Code* (Whittier, Calif.).

Naeim, F., and R. W. Graves
2006 "The Case for Seismic Superiority of Well-Engineered Tall Buildings," *The Structural Design of Tall and Special Buildings* 14, pp. 401–16.

Olsen, K. B., et al.
2006 "Strong Shaking in Los Angeles Expected from Southern San Andreas Earthquake," *Geophysical Research Letters* 33, LO7305, doi: 10.1029/2005GL025472.

Shaw, J. H., et al.
2002 "Puente Hills Blind-Thrust System, Los Angeles, California," *Bulletin of the Seismological Society of America* 92, pp. 2946–60.

Somerville, P. G.
2003 "Magnitude Scaling of the Near Fault Rupture Directivity Pulse," *Physics of the Earth and Planetary Interiors* 137, pp. 201–12.

Somerville, P. G., et al.
1997 "Modification of Empirical Strong Ground Motion Attenuation Relations to Include the Amplitude and Duration Effects of Rupture Directivity," *Seismological Research Letters* 68, pp. 199–222.

———
1999 "Characterizing Earthquake Slip Models for the Prediction of Strong Ground Motion," *Seismological Research Letters* 70, pp. 59–80.

Somerville, P. G., and A. Pitarka
2006 "Differences in Earthquake Source and Ground Motion Characteristics between Surface and Buried Earthquakes," *Proceedings of the Eighth National Conference on Earthquake Engineering* (Oakland, Calif.).

Bibliography

Abrahamson, Norman A., and Walter J. Silva. "Abrahamson and Silva NGA Ground Motion Relations for the Geometric Mean Horizontal Component of Peak and Spectral Ground Motion Parameters," *Earthquake Spectra* 23.

Boore, David M., and Gail M. Atkinson. "Boore-Atkinson NGA Ground Motion Relations for the Geometric Mean Horizontal Component of Peak and Spectral Ground Motion Parameters," *Earthquake Spectra* 23.

Campbell, Kenneth W., and Yousef Bozorgnia. "Campbell-Bozorgnia NGA Ground Motion Relations for the Geometric Mean Horizontal Component of Peak and Spectral Ground Motion Parameters," *Earthquake Spectra* 23.

Chiou, Brian S.-J., and Robert R. Youngs. "Chiou and Youngs PEER-NGA Empirical Ground Motion Model for the Average Horizontal Component of Peak Acceleration and Pseudo-Spectral Acceleration for Spectral Periods of 0.01 to 10 Seconds," *Earthquake Spectra* 23.

Seismic Isolation and Damping Systems: Improving Earthquake Performance of Buildings

Charles A. Kircher **Alex G. Tarics**

Introduction

Recent earthquakes in China and Turkey have once again demonstrated the vulnerability of buildings to strong ground shaking resulting in widespread damage and large loss of life.[1] In the 1995 Kobe Earthquake, over one in every three buildings located close to fault rupture was destroyed. Over six thousand people were killed, and over $100 billion in losses occurred in this event, even though the buildings were primarily of modern construction and designed to withstand earthquakes.

In the United States, we have been spared large losses of life in recent earthquakes—but not large economic losses. In the 1994 Northridge Earthquake, approximately $40 billion of building damage occurred, even though only about twenty people died from the collapse of buildings. The success of our seismic codes in protecting life safety is clear, but costly building repairs and facility downtime have increased interest in improving the earthquake performance of buildings. While improvement in this area has traditionally been considered impractical for many buildings, seismic isolation and damping systems now make enhanced performance both feasible and affordable.

Seismic Performance of Conventional Construction

Construction is one of our oldest activities, yet we have not learned until recently how to erect buildings that will remain essentially undamaged, with their contents safe, after an earthquake. The basic reason may be traced to the fundamental principle of seismic design—prevent building collapse. Traditional design practice ties structures firmly to their foundations and makes them strong enough to resist forces produced by earthquakes. But experience has taught us that this approach generates very high earthquake forces in buildings.

To prevent buildings from collapsing, engineers rely on the ductility of the structural materials. These material stretch before they break; consequently, the buildings may undergo large and sometimes permanent inelastic deformations during a large earthquake. Buildings will not collapse, but they may be so badly damaged that they are unsafe for occupancy and may not be worth repairing. Even if the structural system survives, damage to nonstructural components and contents can affect the operation of the building. Loss of function is unacceptable for essential facilities, such as hospitals, communication and emergency operations centers, and police and fire stations, all of which must remain usable when they are most needed—immediately after an earthquake.

Protecting Buildings with Base Isolation

Over the last fifteen years, seismic engineers have been using a new technology, commonly known as base isolation, as a practical method of protecting buildings from earthquake ground shaking. Rather than tying buildings rigidly to their foundations, seismic isolators support the base of the building while permitting large sideways displacements. The building "floats" on flexible pedestals during earthquakes, greatly reducing the forces in the structure and the shaking felt in the building.

Base isolation has found a wide range of building applications, including protecting the functionality of emergency and critical facilities, protecting the fragile contents of art museums[2] and the sensitive equipment of manufacturing facilities, and protecting the architecture of important historical buildings. Base isolation has been chosen as part of the seismic rehabilitation scheme for a number of existing buildings, since it not only provides superior earthquake protection but also minimizes the need to add new walls or bracing throughout the structure. While California, with about fifty major base-isolated buildings, leads the United States, other states, including Oregon, Washington, Utah, Missouri, and South Carolina, also have base-isolated buildings. Since the 1995 Kobe Earthquake, the number of base-isolated buildings in Japan has increased from about one hundred to over six hundred.

While base-isolated buildings have higher costs of design and initial construction, they can still be economical over the life of the building, considering the reduction in future costs for earthquake repair, loss of function, and business interruption. For new buildings, the added costs of base-isolation design, seismic isolators, flexible couplings for utilities, and other isolation construction could vary between 4 to 8 percent of the total price of the building—about the same as the real estate transaction fee that is assessed each time the building is sold.

Reducing Response with Seismic Damping Systems

Base isolation is not the only strategy for improving the earthquake performance of buildings. Energy-dissipation systems—commonly known as damping systems—have become quite popular in the last five years. Like base isolation fifteen years ago, damping systems are now transitioning from theory to practice and are being used increasingly for seismic projects in the United States and other countries. While damping systems cannot provide the same level of protection as base isolation, they generally have a lower initial cost and offer a design option, with seismic performance somewhere between that of base isolation and conventional building systems.

Seismic damping systems improve building performance by dissipating earthquake shaking energy that would otherwise force the structural system to yield and cause permanent damage. In essence, seismic dampers serve as structural elements that take the punishment of the earthquake. Specifically designed for this task, seismic dampers can survive many cycles of earthquake load without failing or sustaining the level of damage that would otherwise occur to other elements of the structure.

There are many types of seismic dampers, some using friction to dissipate energy, others using fluid viscosity, like automobile shock absorbers. Just as car shocks dampen the response of an automobile, seismic dampers reduce the response of the structure. To be most effective, the dampers are distributed over the height of the building, much like conventional seismic bracing. Unlike conventional bracing, however, they can be used to decrease displacement response without adding stiffness or strength to the building. Seismic rehabilitation projects have found this aspect of dampers to be quite attractive, since it lessens or even precludes costly modifications to existing foundations.

Seismic Codes Supporting the Use of New Technologies

Beginning with the 1994 edition, the National Earthquake Hazards Reduction Program's *NEHRP Recommended Provisions for Seismic Regulations for New Buildings and Other Structures*, commonly referred to simply as the *NEHRP Provisions*, has included design requirements for seismically isolated structures. The Building Seismic Safety Council (BSSC) of the National Institute of Building Sciences (NIBS) prepares the *NEHRP Provisions* for the Federal Emergency Management Agency (FEMA).

The *NEHRP Provisions* provide the role model for seismic codes, such as the new *International Building Code* (*IBC*), which has just been completed and will be used to regulate building construction throughout the United States. These provisions provide engineers and building officials with a common basis for preparing and approving designs of buildings with base isolation systems. Building owners may also be assured that these systems comply with the "code."

Recognizing a need based on the increased interest in damping systems, a new section has just been added to the 2000 edition of the *NEHRP Provisions* for the design of buildings that incorporate seismic dampers. The new provisions are currently being reviewed by the BSSC membership and will likely become part of the 2003 edition of the *IBC*.[3]

Conclusion

The acceptance and incorporation of new technologies like base isolation and damping systems in our building practices is a very complex process that requires the transfer of technology from researchers and specialists to practicing structural engineers. Seismic codes play an important role in technology transfer by providing practicing engineers with established criteria and methods for design.

Information on new technologies must also be transferred to architects, who advise building owners on building design, and directly to building owners and managers. Representatives of federal, state, and local government agencies also must be informed, because they make the final decisions on what kind of earthquake protection should be provided for our public buildings and facilities.

Much of the time and effort needed to transfer this information is already behind us. The technology has been adequately developed and is available to contribute significantly to solving the problem of improving building performance—reducing damage and keeping essential buildings and facilities functional after an earthquake.

Notes

1. This paper was published as "Seismic Isolation and Damping Systems—Improving Earthquake Performance of Buildings," *Building Sciences* 24 (Aug. 2000).

2. Two major art museums in San Francisco have chosen seismic isolation for protection of their buildings and collections—the Asian Art Museum (see Donna Strahan, "Seismic Mitigation at the Asian Art Museum," in the present volume) and the new de Young Museum in Golden Gate Park.

3. Design provisions for seismic isolation and damping systems are now incorporated into the 2003 and 2006 editions of the *International Building Code,* as well as into national standards published by the American Society of Civil Engineers (ASCE): ASCE/SEI 7-05, *Minimum Design Loads for Buildings and Other Structures*; and ASCE/SEI 41-06, *Seismic Rehabilitation of Existing Buildings.*

Developments in Seismic Protection of Historic Structures

William S. Ginell

Introduction

The vulnerability of many historic buildings to damage or collapse during earthquakes poses substantial life safety risks to building occupants and also to culturally significant artifacts contained within the structures. Irreplaceable historic properties can be affected and within instants, meticulously conserved objects in collections can be compromised or destroyed. To minimize loss of life and to protect the historic building's fabric and its contents from catastrophic damage, some form of structural modification, usually known as seismic retrofitting, needs to be implemented as an alternative to the costly and time-consuming effort of replacing the entire building.

The problem of protection of vernacular structures from damage by disastrous earthquakes was recognized in ancient times. By trial and error, structural techniques were developed that resulted in only minimal or repairable damage to buildings rather than complete collapse accompanied by significant loss of life. These techniques included: construction of buildings with thick walls, low ceiling heights, square floor plans, small penetrations for windows and doors, and location of openings away from corners. A major innovation was the inclusion of ladder-type wooden or solid timber ring beams or belts located at regular elevations within the walls. The building shown in figure 1 is located in a highly seismic area of the Republic of Macedonia, and its survival is due in part to the timber ring beams in the mud brick (also called adobe brick) walls.

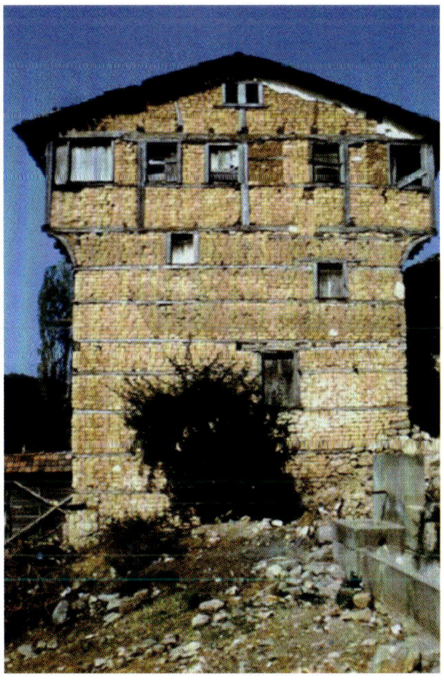

Figure 1
Multistory adobe structure in Macedonia. The lowest level is constructed of stone and the next level of adobe bricks. The top level consists of a wood frame with an adobe brick infill.

Ring beams of this type increase the ductility of a structure, and they have been proven to be effective in preventing wall collapse. They can still be seen in mud brick and stone-walled structures ranging from southern Europe to highly seismic areas in eastern Asia. However, with time, these construction details were gradually forgotten or were modified because timber was expensive and the earthquake occurrence frequency in these regions may have diminished to the degree that this type of protection was no longer needed. Furthermore, the use of newer construction materials such as concrete or concrete blocks became common, and these seemed to provide adequate, inexpensive, and safe structures. Adobe walls in particular, as well as other unreinforced masonry structures, no longer were constructed with wooden ring beams, a fact that may have contributed to the widespread destruction of historic vernacular buildings and villages during periods of intense seismic activity.

In recent years, considerable progress has been made toward understanding the engineering behavior of buildings during earthquakes. Unreinforced masonry buildings are universally recognized as highly vulnerable, and in California at least, state law requires that the hazard posed by such existing buildings be reduced either by retrofitting or, if this is not carried out, by demolition (California State Historical Building Safety Code Board 1999). Seismic retrofitting should be included in plans for the protection of historic buildings, and such techniques fall within the emerging field of conservation engineering. It has been stated that conservation engineering "is distinguished by application of rational analysis and design techniques to buildings containing structural elements that cannot be analyzed using current methods or were not designed using currently available materials. Such elements, defined as archaic, obsolete, or undesigned, are marks of a 'structurally historic' building and require specialized methods. Common methods of structural analysis and design have evolved for use in new construction and are not well suited for use in conservation work" (Friedman 2001).

In addition to the technical reasons for difficulties in incorporating retrofitting technologies into historic buildings, ethical considerations must also be taken into account. Modification or loss of authenticity is usually the result of physical interventions that do not recognize or are applied regardless of potential damage or loss to architectural details, wall paintings, decorations, or other unique characteristics of the building. Some structural modifications to the building could change internal or external load paths that could lead to loss of stability during a future earthquake. Therefore, the design of an effective structural retrofit requires recognition of the need for both preservation of authenticity values and the application of sound engineering judgment. To help assure these aspects, cooperation is needed among architectural historians, structural engineers, architects, conservators, materials specialists, and local officials who are familiar with building codes. In most cases, building codes do not address obsolete materials, construction methods, or the current state of degradation of historic structures and are therefore not generally applicable. It is becoming standard practice that when retrofitting plans are submitted to building officials for approval, structural engineers request waivers of code compliance on the basis of having demonstrated sound technical judgment in formulating the design.

To illustrate several recent methodologies used for retrofitting structures, this paper presents two research case studies and a related project in which the Getty Conservation Institute (GCI) was involved. These examples illustrate the practical problems of building preservation and are important to the protection of museum collections. Although the artifacts that make up a collection may themselves be mounted or stored so that no damage would be incurred during the postulated design earthquake, collapse of the building could very easily negate the seismic mitigation features of the mounts. Accordingly, assuring the seismic stability of a building is a requirement for protection of collections in seismically susceptible structures.

Retrofitting of Adobe Structures

Vernacular earthen buildings, such as those constructed of mud bricks or rammed earth, continue to be in use throughout the world. Recent severe earthquakes in Turkey, Iran, India, Indonesia, Latin America, the southwestern United States, and other countries have devastated many such structures, usually with an attendant loss of life. Earthen materials have negligible tensile properties, and under seismic loading, walls crack and overturn and roofs collapse.

The terms *mud brick* and *adobe brick* are used interchangeably to describe the building blocks that can be used to form an earthen wall. However, not all types of mud can be used—only mud of certain compositions is suitable for making bricks that can be used to build walls. Another type of adobe wall is sometimes built by casting mud between vertical wall forms. If the mud has sufficient plasticity and cohesion, it can be tamped in place between the forms to yield a solid structure that is self-supporting after the forms are removed. These types of structures are also called adobe walls or, more descriptively, rammed earth walls. The seismic retrofit methods discussed in this paper are applicable to both types of earthen walls.

Particularly vulnerable are thin adobe walls and walls with a slenderness (height-to-thickness) ratio greater than 6. Earthquake forces parallel to the plane of the wall (in-plane forces) produce tensile stresses and diagonal cracking; forces perpendicular to the plane of the wall cause flexural stresses, cracking, rocking, and eventually overturning. Both types of wall failure can result in roof collapse. The adobe wall fabrics of many historic structures are integral parts of the cultural history of the building, and wall decorations, murals, paintings, and wooden architectural details are also features that require protection. In the event that retrofitting of the building is being considered, attention must be given to the protection of such features.

In the early 1990s, the GCI embarked on a program, the Getty Seismic Adobe Project (GSAP), to devise and test retrofitting technologies that would minimize structural damage and yet be compatible with the preservation of the historical and cultural aspects of adobe buildings. In all the studies carried out, technologies were chosen that would be minimally invasive and reversible, if possible.

Typically, conventional retrofitting technology seeks to improve the *strength* of buildings so that they can continue to react elastically to seismic shaking. However, having reached the elastic limit, failure occurs as a result of cracks and serious loss of structural strength. In the case of earthen buildings, the elastic limit is exceeded at very low seismic input accelerations, and cracking takes place almost immediately. The fundamental premise behind the GSAP technology was to improve the *stability* of the building rather than its strength, so that, although cracked and inelastic, the walls remain standing and continue to support the roof. It was felt that with stability-based measures, the seismic performance of unreinforced adobe structures could be greatly improved by the use of minor restraints and continuity elements that inhibit the relative displacement and loss of cracked wall sections, which constitute the principal mode of failure. The slenderness ratio and absolute wall thickness would determine the most appropriate type of retrofitting measures among those tested.

Initial tests were performed on a shake table at Stanford University on 1:5 scale models of both as-built and retrofitted adobe structures (Tolles et al. 2000). Modeling theory and the laws of similitude were used throughout the test program to design the test structures; this process resulted in a complete set of correlation functions that defined the model-prototype relationship. The retrofitting measures used consisted of: vertical and horizontal straps or cables attached to the external and internal surfaces of the walls; wooden bond beams anchored into the tops of the walls; an attic floor partial diaphragm attached to the roof and the bond beam; and vertical center-core rods incorporated within the out-of-plane walls and at other selected locations. Center-core rods were also used where the use of straps could damage wall decorations. These measures were found to be highly effective because the relatively large cracked wall sections that were formed during shaking were held in place by the straps and cores and thereby continued to support the roof and prevent total collapse. In practice, the straps or cables would be recessed below the surface of the wall and covered with adobe mud so that no retrofitting measures would be visible. This procedure is acceptable since periodic resurfacing of the outer walls of vernacular earthen buildings and the cosmetic grouting of superficial cracks are still common practices—for example, in the American Southwest.

The tests showed that application of stability measures, straps, and center cores was highly effective in preventing catastrophic damage to the models. Figure 2a shows the unretrofitted adobe model, and Figure 2b shows the retrofitted model after shaking. Increasing the wall thickness and decreasing the slenderness ratio from 11 to 5 also had positive effects on seismic performance.

Further tests were carried out on two large-scale (1:2) models on a shake table in Skopje, Republic of Macedonia. One model was unretrofitted, and the other was retrofitted during construction. These tests were carried out to determine if gravity loading effects that were unobservable on the 1:5 scale

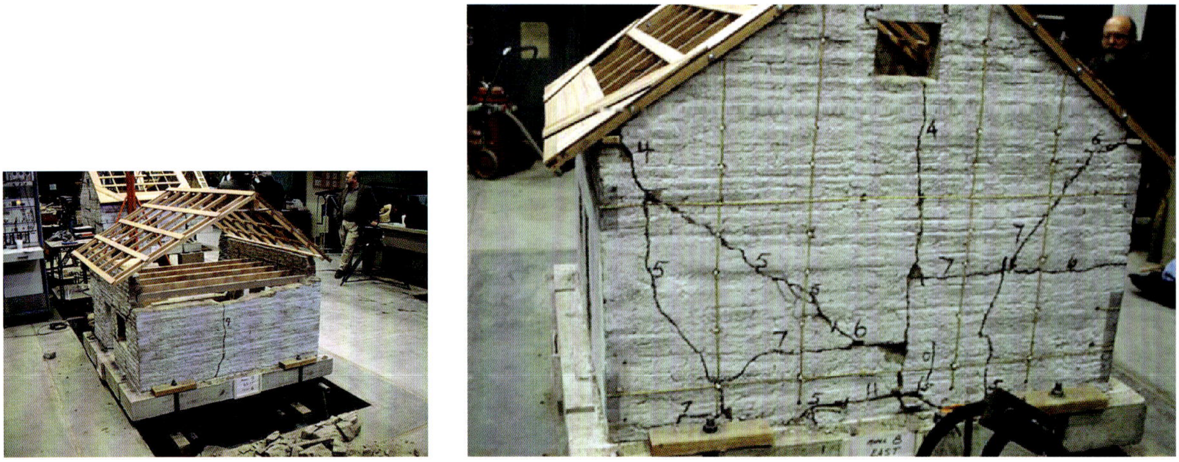

(a) (b)

Figures 2a and 2b
Unretrofitted 1:5 scale model adobe building after shaking (a), and retrofitted 1:5 scale model adobe building after shaking (b).

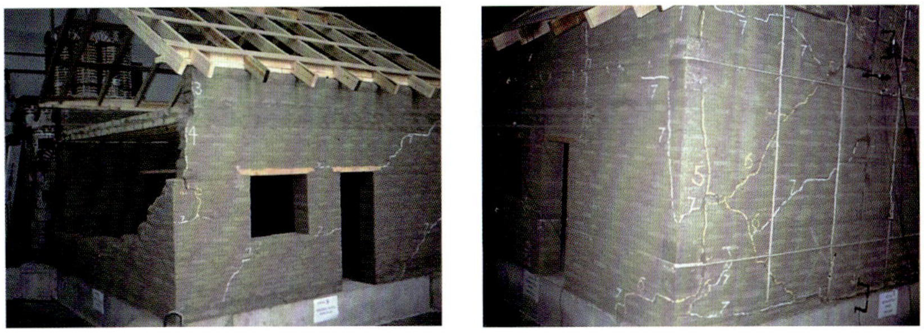

(a) (b)

Figures 3a and 3b
Unretrofitted 1:2 scale model adobe building after shaking (a), and retrofitted 1:2 scale model adobe building after shaking (b).

model walls would be important on a model more nearly full size. Test results indicated that the behavior of the large-scale models was very much the same as that of the smaller-scale models and that gravity loading had only a slight effect on the failure mode and stability of the model (Ginell et al. 2000). Figure 3a shows the complete collapse of the gable end wall of the unretrofitted model after shaking, while figure 3b shows the corresponding wall of the retrofitted model. Although damaged, this wall remained upright and was repairable.

The stability-based concept and technologies for retrofitting historic adobe structures that were developed at the GCI have received wide acceptance throughout the architectural conservation community. A summary of the planning and engineering guidelines for retrofitting historic adobe structures is given in Tolles, Kimbro, and Ginell (2002).

Byzantine Church Retrofitting

Mud or adobe brick and other adobe buildings are usually unreinforced structures. Many stone and masonry brick buildings were also constructed without reinforcement and are also vulnerable to seismic damage. For example, stone churches built during the Byzantine period in Europe and the Near East have been severely damaged by the frequent earthquakes in those regions. Many have survived although damaged, because of the inclusion of circumferential timber ring beams or belts located within the walls—the same seismic mitigation architectural elements that were included in old adobe structures in the same regions. Unfortunately, decay of the timbers, resulting in the formation of voids in the walls, has reduced the ability of some of these churches to resist seismic shaking. Many of the churches have been completely lost, and of those that remain, few original towers and domes still exist.

To improve the chances of survival of these ancient Byzantine churches, the GCI, in collaboration with the Institute of Earthquake Engineering and Engineering Seismology (IZIIS) in Skopje, embarked on a project to devise and test methodologies for retrofitting the churches using minimal intervention techniques that would preserve the unique cultural and historic features of these increasingly sensitive structures (Gavrilovic, Sendova, and Ginell 1999).

Experimental and analytical investigations were performed to verify an original procedure for the repair and seismic strengthening of churches of this type in Macedonia. After a detailed study of the condition and structural design of about fifty surviving churches, the St. Nikita Church located in the village of Banjani was selected as a prototype to be representative of churches in the Balkan region (fig. 4).

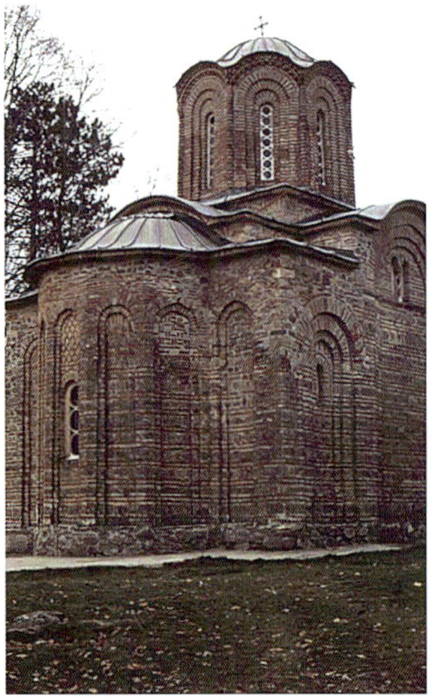

Figure 4
St. Nikita Church in the village of Banjani, Macedonia. This church was selected as a prototype for the test church model.

The church, a single-dome structure with a cross inscribed in a rectangular plan, consists of massive perimeter walls in both orthogonal directions and internal columns supporting vaulted roof elements and the central dome. The original church still contains frescoes produced in the fourteenth century (fig. 5), and the retrofits were designed to protect these important historic religious artifacts as well as both the structural and the architectural features of the church. Investigations showed that the original timber belts that had been installed to sustain tensile forces had, to a large extent, disintegrated; this loss was the principal cause of the decrease in stability and damage to the church that had occurred during previous earthquakes.

A model of the church was built to a scale of 1: 2.75; the model was constructed of the same stone, brick, and mortar as the church itself (fig. 6). Analytical studies were performed to determine the physical, mechanical, and chemical characteristics of the materials used in the original church, in order to specify components that would ensure a truly accurate model—not only dimensionally accurate but also structurally accurate. Of special importance were the degraded properties of the existing lime mortar and stone. The seismic hazard parameters of the prototype church region were determined from geological and geotechnical studies, and the maximum expected accelerations and seismic intensities for certain earthquake return periods were estimated. These data were required to be able to achieve an accurate simulation of the expected environment during the shake table tests to be performed on the models.

The seismic stability of the prototype church and its response to earthquake vibrations were determined, in order to estimate the ultimate bearing and deformability capacity of the structure for different intensities and types of seismic excitations. This determination was achieved by measuring the response of the actual church to ambient vibrations caused by the wind and by simulated forced vibrations produced by a mechanical vibrator attached to the building during the testing. These measurements resulted in determination of the natural frequencies, resonances, and mode shapes of the church in the north-south and east-west directions. The values obtained for the model were closely comparable to those of the prototype, a finding that assured an accurate simulation. Finally, a mathematical model of the church was constructed based on the measurements. This step was followed by the development of possible engineer-

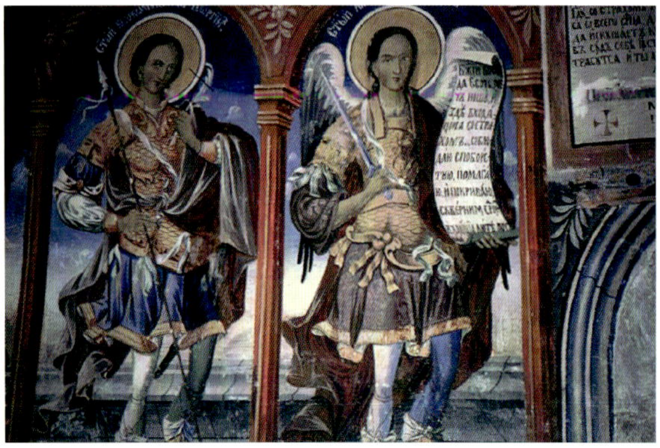

Figure 5
Fresco in St. Nikita Church.

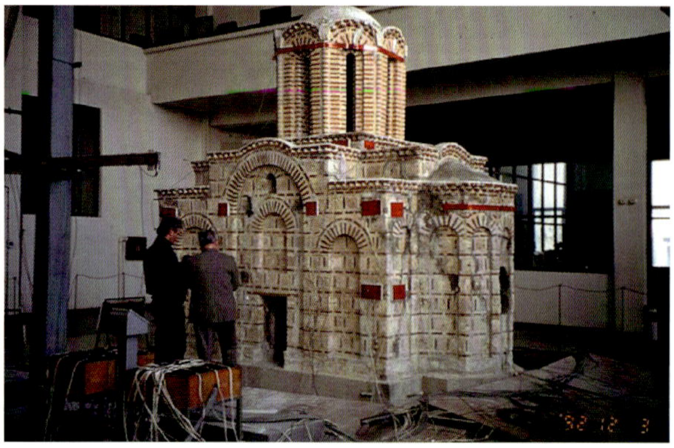

Figure 6
Model of St. Nikita Church (scale 1:2.75).

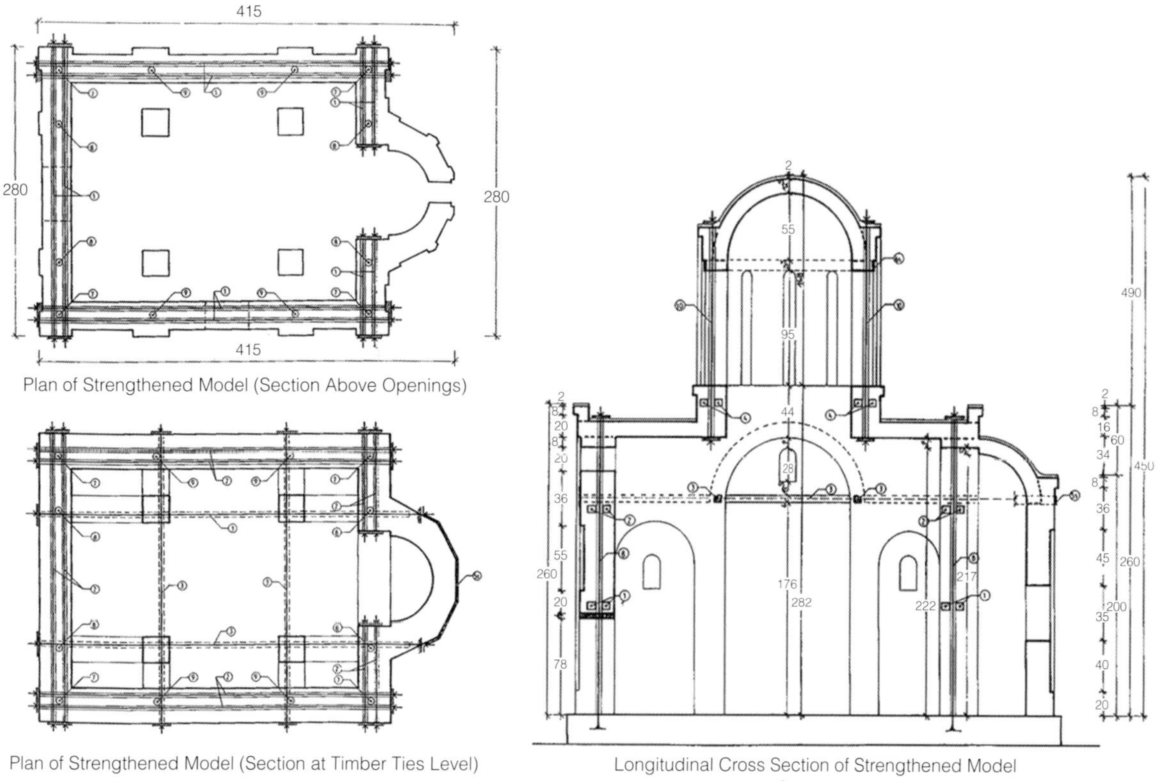

Figure 7
Plan views and an elevation section of the St. Nikita Church model, showing the placement of horizontal and vertical retrofitting rods.

ing concepts for the repair and strengthening of the church. It was decided that incorporating horizontal and vertical steel tie belts in the walls and filling the void spaces around the ties with a compatible grout material would serve to increase the bearing capacity and deformability of the existing structure. Where possible, the steel ties would be placed in the wall regions that had previously been occupied by the timber belts (fig. 7).

The model church was constructed on a concrete slab that could be mounted on the 5 m (16 ft., 4 in.) square, computer-controlled shake table at IZIIS. The scaled time-history properties of three different types of real earthquakes were incorporated into the computer program, so that both local and distant earthquakes would be considered. The earthquake time histories were applied in the north-south direction of the model by stepwise increases in the shake table maximum acceleration amplitude from 0.05 to 0.43 g. The model was fully instrumented to measure accelerations, deflections, and various other forces.

Videos of the behavior of the model were taken from all angles during the shaking, and still photographs of the development of damage were also obtained after each test. It was found that the model behaved elastically, and no cracking occurred up to an acceleration of 0.17 g, after which cracks formed; at 0.43 g, damage to the building approached complete failure.

The model was then repaired by filling the cracks with a special lime mortar injected under controlled pressure. The horizontal strengthening steel belt rod elements were installed in the walls in channels incorporated during construction. The rod elements were then prestressed. Vertical steel reinforcing rod elements were installed at the ends of the walls and around wall openings. These rod elements served to increase ductility and bending resistance. Injection grout mortars for crack repair and for filling the volumes around the steel reinforcement rods were different in both chemical composition and grain size distribution, to allow for differences in requirements at the two types of locations.

After a suitable retrofitting and curing time, the model was then subjected to the same series of tests as was performed previously on the original model. The results showed that the retrofitted model remained elastic up to an acceleration level of 0.40 g, at which point the first crack appeared. At an input acceleration of 0.54 g, corresponding to an earthquake with an expected return period of 1000 years, the model suffered only some minor repairable structural damage. Cracks continued to develop, and at the maximum attainable level of table acceleration (0.75 g), the structure behaved very nonlinearly but was still far from failure. The final damage level of the strengthened model was considerably lower than that of the original, unretrofitted model. However, it was noted that the failure mechanism changed from the separation of bearing walls to the appearance of diagonal cracking of the lower walls due to shear stresses.

This was the first shake table test of a large-scale model of a historical building that demonstrated a successful methodology for conservation of the building itself, as well as for preservation of its interior and exterior architectural and decorative features (Gavrilovic et al. 2004).

Base Isolation Seismic Retrofitting

The third example of modern engineering strategies for the upgrading of historic buildings that are seismically unsafe involves a method known as base isolation. Rather than preventing cracked walls from overturning (as was done for the adobes) or strengthening massive walls to improve their bearing capacity and elastic limits (Byzantine churches), this method seeks to prevent damage by isolating the structure from the seismic source by changing the frequency response of the building, so that resonances with the damaging earthquake frequencies do not occur, and energy transfer into the structure is minimized. Application of this technology has been used mostly on new construction throughout the world, but it was first implemented in the United States in an existing building, the historic City and County Building, a structure in Salt Lake City, Utah, that was being considered for seismic upgrading (Bailey and Allen 1988).

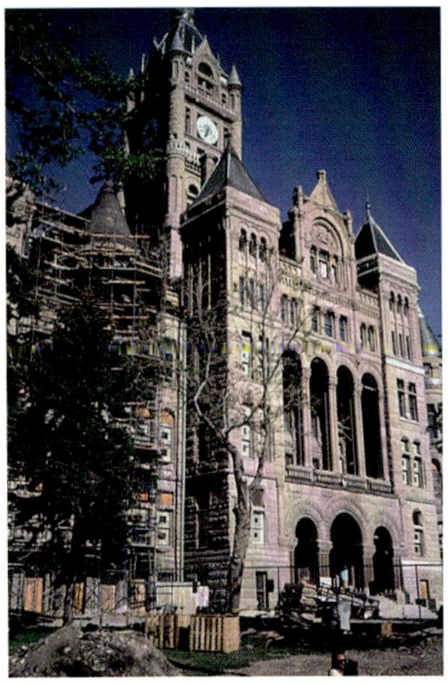

Figure 8
City and County Building in Salt Lake City, Utah.

The style of the building's architecture is Richardson Romanesque (fig. 8). It is a large, five-story, unreinforced masonry and stone structure with a 76 m (249 ft.) central clock tower built in the early 1890s. Salt Lake City is an area of relatively frequent seismicity, and the city administration was faced with three possible choices for addressing the problems of obsolescence and life safety: (1) tear the building down and rebuild, thus destroying an important cultural asset; (2) retrofit the building by removing the entire internal structure and erecting a modern reinforced concrete and steel framework within the stone shell that would be strong enough to resist an expected earthquake (resulting in the loss of the ornate original interiors installed about a century earlier); or (3), use the new concept of base isolation to preserve most of the historic features of the original building and restrict changes to those elements occurring only at the lower supporting structure of the building. The choice was made to proceed with the base isolation option.

To achieve base isolation, the building must be completely detached from direct contact with the ground and mounted on devices that decouple the structure from the horizontal vibrational components of the earthquake. Isolation devices for this purpose are themselves simple and inert, except when activated. The isolators used in the City and County Building consisted of a stack of alternating steel and rubber plates. The isolator is stiff in the vertical direction and is capable of supporting heavy loads. However, it is flexible in the horizontal direction and can withstand shear forces. A vertical lead core through the center of the plates acts as a damper and prevents any relative movement of the plates until a predicted shear force (produced by earthquake shaking) allows movement of the uppermost plate with respect to the lowest plate.

Each of the 447 isolator units was 43 cm (17 in.) square by 38 cm (15 in.) high, and the units were installed at appropriate support locations under the building (fig. 9). The upper parts of the isolators were installed on the bottom of the stiffened base of the building, and the lower parts of the isolators were attached to the foundation of the building, which is in contact with the ground. During an earthquake, the ground moves, but there is a delay in the transfer of energy through the isolator as the rubber shears. The vibration period of the building is shifted and becomes lower than the damaging dominant frequencies of the quake, the energy difference being absorbed by the rubber. The lead core acts as a fuse and does not allow building movements that could occur from vibration sources, such as the wind or traffic, below the preset minimal limits representative of a destructive earthquake.

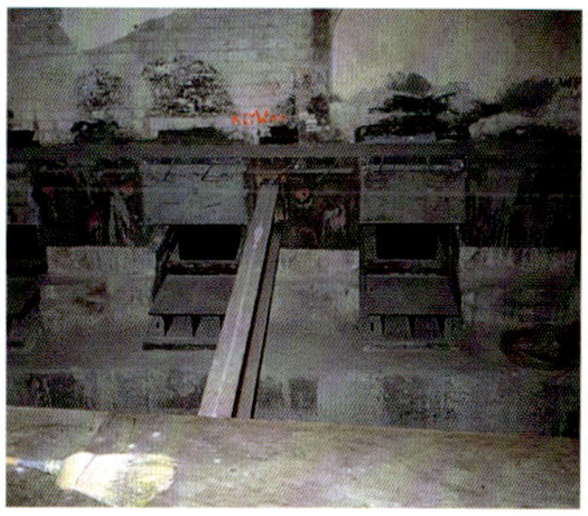

Figure 9
Multiple base isolators installed under the reinforced ground floor of the City and County Building.

The structural changes to the building that were required for the installation of the isolators included providing a strong, continuous lower floor and providing reinforced concrete mounting structures for the isolators below the floor. To install the isolators between the footings and the lower floor so that the building rested only on the isolation devices, it was necessary to cut through the existing concrete foundation, which was done with a diamond-impregnated cable saw (fig. 10). A 30.48 cm (12 in.) wide, covered trench was provided all around the exterior of the building to allow for the translation of the building itself with respect to its footings during an earthquake.

Extensive engineering analyses were performed prior to construction, including the development of mathematical models and predictions of anticipated building movements during the expected earthquakes. Since completion of the project, low-intensity earthquakes have occurred over time, and the building has responded as predicted and without damage to the structure or its contents. Seismic isolation as a methodology for the protection of buildings is now widespread, and a large number of isolator designs have been devised, tested, and installed. The same principles of base isolation have also been applied to the protection of artifacts in collections.

Seismic Isolation of the Byzantine Church Model

After completion of the repairs, strengthening, and retesting of the Byzantine church model, it was decided to determine if base isolation is a viable retrofitting technology for historic structures that might be damaged when subjected to necessary interventions that could indirectly affect fresco wall paintings. The advantage of base isolation is that construction work is

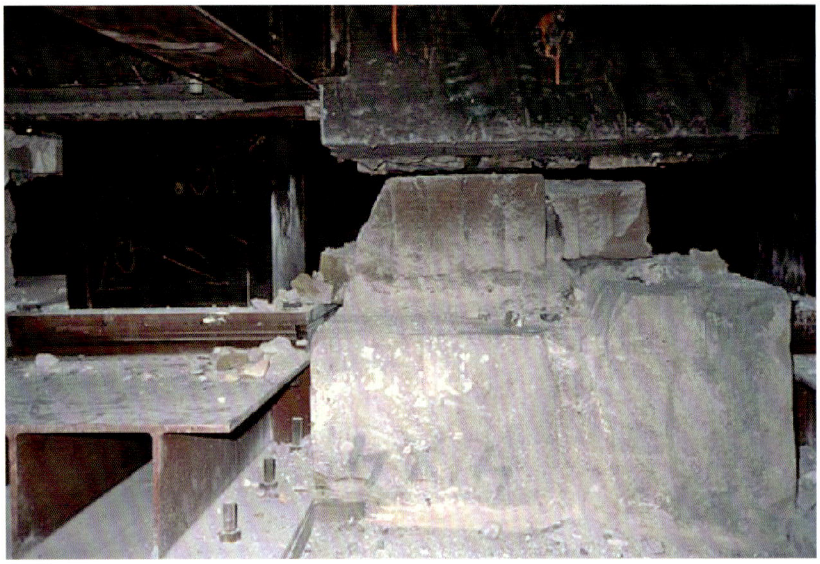

Figure 10
A single isolator supporting one section of the building.

performed only at the foundation level while other conservation efforts can be carried out simultaneously on the existing upper structure. Frescoes can thus be protected from possible damage during construction and can be conserved independently. A disadvantage of base isolation is that the traditional support structure of the building must be modified through an approach based on modern, technological innovation. Any archaeological evidence of cultural heritage at the base of the structure could be lost as a result of the irreversible modifications required by base isolation. This possibility represents a compromise with one of the foremost tenets of current conservation ethics—the retention of the original characteristics and fabric of a structure.

To implement the tests, modifications to the base of the church model were made to accommodate eight rubber and steel isolators interposed between the concrete support slab and a steel structure that was built to connect the isolators and model with the shake table (fig. 11a). In figure 11b, the model is shown mounted on the isolators. Steel stopper elements designed to limit the maximum translation of the building were also installed. The same series of earthquake simulation accelerations was activated as had been carried out previously. Figure 11c shows the shearing of the rubber components of the isolator that occurred during the test.

The results indicated a 50%–60% decrease in the acceleration imparted to the model when compared with that of the strengthened model, resulting in much smaller horizontal displacements, especially at the dome level. Up to an input acceleration of 0.60 g, the base-isolated model behaved like a rigid body, and no visible cracks were formed. Because the church model had been strengthened before the base isolation tests, it cannot be said that base isolation alone would have yielded results as spectacularly effective as were demonstrated. Yet the reduced response was impressive, and base isolation appears to be a viable technique for historic building retrofitting (Gavrilovic and Sendova 2001).

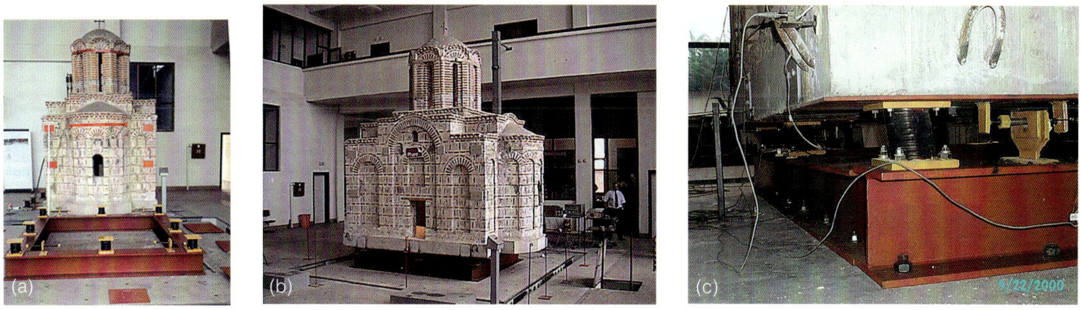

Figures 11a–c
Church model before its installation on the isolators (a); the church model mounted on the isolators (b); and shearing of the rubber components of the isolators during the test (c).

Conclusion

Effective engineering technologies now exist for safeguarding historic structures and museums by providing protection for collections from the devastating effects produced by earthquakes. The friction pendulum system (FPS), for example, is a design that was originally devised for buildings but which has been adapted for use in collections (see Zayas, Low, and Mahin 1987).

The structures of house museums and larger buildings that serve to house and protect collections can be altered in accordance with ethical conservation principles, although often at some considerable expense. There are no standard designs that would solve all problems, and each prospective retrofitting of a building must involve a detailed engineering study to establish objectives, as well as require the formulation of a plan that is unique to the structure. Protection of a collection of artifacts from seismic damage must rely not only on the effectiveness of seismic mitigation mounts but also on the integrity of the building in which the collection is housed.

References

Bailey, J. S., and E. W. Allen
1988 "Seismic Isolation Retrofitting Salt Lake City and County Building," *Journal of Preservation Technology* 20, no. 2, pp. 32–44.

California State Historical Building Safety Code Board
1999 *State Historical Building Code (SHBC): Part 8, Title 24, California Code of Regulations,* rev. June 1990 (Sacramento, Calif.).

Friedman, D.
2001 "Methodology of Conservation Engineering," *Journal of Architectural Conservation* 7, no. 2, pp. 49–63.

Gavrilovic, P., V. Sendova, and W. S. Ginell
1999 "Seismic Strengthening and Repair of Byzantine Churches," *Journal of Earthquake Engineering* 3, no. 2, pp. 199–235.

Gavrilovic, P., and V. Sendova
2001 *Protection of Byzantine Churches against Earthquakes Using Seismic Isolation,* Report IZIIS 2001-59 (Skopje, Macedonia).

Gavrilovic, P., W. S. Ginell, V. Sendova, and L. Sumanov
2004 *Conservation and Seismic Strengthening of Byzantine Churches in Macedonia,* Getty Conservation Institute Scientific Reports (Los Angeles).

Ginell, W. S., E. L. Tolles, P. Gavrilovic, and V. Sendova
2000 "Seismic Shaking Table Tests of Retrofits for Large-Scale Model Adobe Structures," *Proceedings of the Eighth International Conference on the Study and Conservation of Earthen Architecture* (London).

Tolles, E. L., E. E. Kimbro, F. A. Webster, and W. S. Ginell
2000 *Seismic Stabilization of Historic Adobe Structures: Final Report,* Getty Conservation Institute Scientific Reports (Los Angeles).

Tolles, E. L., E. E. Kimbro, and W. S. Ginell
2002 *Planning and Engineering Guidelines for the Seismic Retrofitting of Historic Adobe Structures,* Getty Conservation Institute Scientific Reports (Los Angeles).

Zayas, V. A., S. S. Low, and S. A. Mahin
1987 *The FPS Earthquake Resisting System—Experimental Report,* Report No. UCB/EERC 87/01 (Berkeley, Calif.).

Seismic Protection of Museums and Museum Collections in India

O. P. Agrawal

Introduction

January 26, 2001, was to be a day of celebration. On that day each year, India celebrates its 1950 declaration of the sovereign Republic of India. It is a day known for rejoicing. On this day there were to be parades of schoolchildren in the morning; flag hoisting ceremonies; and decorated tableaux depicting Indian life, culture, arts, and achievements in every city. Preparations for all the celebrations had been completed in advance. And then, suddenly, at exactly 8:46 A.M., everything collapsed. News came that there had been a severe earthquake in Gujarat, a state in western India. Thousands of houses had collapsed and rendered hundreds of thousands of people homeless and without any support. News emerged that a large number of heritage properties and museums in the area had also been destroyed.

S. K. Misra, Chairman of the Indian National Trust for Art and Cultural Heritage (INTACH), as well as all of us connected with INTACH were very much concerned. A team consisting of conservation architects, art conservators, and natural heritage experts was immediately dispatched to visit Gujarat and prepare an assessment report. The team, which was led by Divay Gupta, Director (Projects) in the Architectural Heritage Division of INTACH, discovered that the Gujarat Earthquake not only destroyed thousands of tall buildings, offices, and houses in towns and villages, but it had also taken a severe toll on the rich architectural and cultural heritage of the region. The team found that the earthquake had affected every type of building, religious as well as secular structures, art objects, and decorative doors and windows. INTACH immediately set up a task force to begin the restoration of some of Gujarat's cultural heritage. The team collaborated with both governmental and nongovernmental agencies in the rehabilitation efforts.

While other agencies were directing their energy toward new structures such as houses, schools, hospitals, and offices, the main concern of INTACH was the restoration of historic architecture, art, and general cultural heritage. It is very possible that without the efforts of INTACH, little or no attention would have been paid to cultural heritage.

The Challenges of Gujarat

Salvage of damaged objects was a significant challenge in Gujurat. Unfortunately, the buildings that held the collections were also those that were least resistant to earthquakes. There were no post-earthquake preparations of any type, nor were there any response plans to depend upon in the immediate aftermath of the devastating earthquake.

One of the greatest difficulties facing the INTACH team during the salvage operation was that, in many instances, the broken objects were buried in the debris of damaged or collapsed buildings. It proved extremely difficult to sort out, or even distinguish, the broken pieces of artifacts from the surrounding rubble. Although we at INTACH sometimes succeeded, much was inevitably lost. In the Bhuj area, where there was a concentrated cluster of museums and palaces, the delicate art objects in those collections were turned to small pieces of rubble, even to dust. No documentation of the objects was available —probably none had existed. But even if it had existed, it too was destroyed or buried under the debris. Restoration and even partial reconstruction were a very difficult proposition, if not a totally impossible one.

At this stage, INTACH focused on post-earthquake repair and restoration. Gujurat had no program of damage mitigation in place before the earthquake to save the cultural heritage and art objects from the effects of the event, and in that sense it was not exceptional in India.

Frequency of Earthquakes in India

The Gujarat Earthquake of 2001 was not the first devastating earthquake in India. Seismic events occur every three or four years in some part or another of India (appendix 1). More recently, for example, there was an earthquake in Jammu and Kashmir, in the northernmost part of India. As in Gujarat, there was vast devastation. Houses collapsed, a large number of people were killed, and a large number of the survivors became homeless. There was significant damage to the artistic and cultural properties of the region as well. As in the case of Gujarat, INTACH Chairman S. K. Misra dispatched a team led by Jahnwij Sharma, Director (Projects) of the Architectural Heritage division of INTACH. An assessment report was prepared; the findings were similar to those for Gujarat. Broken art objects were buried under the debris of the buildings, many of which were historic structures, and it was often not possible to recover the objects for restoration. Again, as in Gujurat, there were very few pre-earthquake precautions in place that might have saved many of the objects and minimized damage to the buildings. Lack of any response or recovery plan made the coordination of salvage efforts difficult and chaotic. Although INTACH was able to inspect and carry out limited salvage operations in many historic structures, the agency has its limitations, and it was not possible to inspect all the historical buildings. In the months that followed, we did try to restore as many buildings as possible, but there were far too many to address them all. The national government's Archaeological Survey of India and the state government's Department of Archaeology were more concerned with the buildings considered important enough to be on the list of protected structures. But there are thousands of other historic buildings making up the fabric of Jammu and Kashmir. They too are important and unique, although they might not be included on the protected list. There was no other agency besides INTACH to look after them.

The issue is not that there had been no damage to architectural and cultural heritage in the earthquakes prior to the instances mentioned. It is only that there had been no organization like INTACH, which was established in

1984, to take up the cause of cultural property. Therefore, much of the previous damage has gone unnoticed and probably uncared for. If a historically important old building were damaged, it must have been demolished and a new one erected in its place. Even in Gujarat, our team found it very difficult to convince the municipal authorities that historical buildings needed to be restored in the same manner that cultural property should be restored. Their approach was to demolish the damaged buildings and to construct new ones in their place.

Precautions in Museums

In museums, palaces, temples, churches, and other repositories, where art objects of various types are frequently displayed and stored, precautions for the safety of the buildings, as well as of the art objects stored and displayed inside them, must be taken. It is important to know what is needed and what has already been done. To evaluate the status of earthquake preparedness in the museums of India, INTACH undertook a national survey focused on museum preparedness. Our teams visited 125 museums of different categories, including museums supported by national and local government agencies, as well as university museums and privately funded museums. We selected the sites carefully to ensure a good geographical representation. We wanted to find out whether museum authorities were aware of the harm that can be done to the objects in the absence of mitigation efforts. We also wanted to know whether any one had received any training in the area of damage mitigation. The results of the survey were not unexpected. Awareness regarding seismic hazards was completely lacking. No one knew about the type of precautions that could be taken or should be taken. No one had received any training whatsoever on these topics. In fact, we found that in the entire country, no formal training focused on earthquake preparedness in museums existed.

Our next question was whether the museum directors thought that training in the safeguarding of museums and museum collections against earthquake damage was necessary. People were not clear as to what type of training could be offered, but many agreed it would be helpful.

Seismic Design Codes

For construction of new buildings, several countries have developed seismic design codes. In India there are several Indian seismic codes formulated by the Bureau of Indian Standards. A list of these codes is given in appendix 2. Such codes are extremely important for guiding architects in the planning, design, and construction of earthquake-resistant buildings. Properties of such buildings have been described in detail by several authors (see Bibliography). They include the capacity to withstand significant deformation under severe shaking during the earthquake.

It will be seen from the list of the Indian seismic codes in appendix 2 that there are different codes for different types of buildings and structures, including mud houses, bridges, and multistory buildings. None, however, have been specifically developed for museums. Because of the delicate nature of the objects a museum structure holds, this category of buildings should receive a treatment different from one suitable for other types of buildings. There is clearly a need to develop such a seismic code for museums, whether they are new structures or older, retrofitted ones.

Studies of Museums

Around the world, safety studies have looked at a variety of structures used for different purposes. For example a great deal of effort has been applied to keep schools safe during earthquakes. This very important work has been conducted by experts with a full knowledge of the specific functions and needs of educational facilities. A similar project must be undertaken to ensure safety in museum structures. Museum officials and the architects they contract to design and build (or renovate) museum structures would benefit from a building code detailing different types of precautions specific to the function and needs of museums.

In India a seismic zone map has been prepared. It has proven helpful in the design of new buildings and in making buildings earthquake resistant. Such a map is clearly pertinent to the design of museums as well as other structures. It must be mentioned, however, that these maps are constantly revised and updated, and one must be careful to reference the most current map.

Training

It is clear that training museum professionals in the area of earthquake damage mitigation is a necessary goal. A proper syllabus and curriculum for such training must be developed, and it should focus on two specific groups: architects and engineers (who design museum buildings and construct them), and directors, curators, and collections care professionals, including restorers and conservators (who form and maintain the collections). This training would support the design of buildings that are resistant to earthquakes and support the adaptation of damage mitigation measures to the display and storage of collections. Other types of training should involve systems for display, including the design of showcases for which materials and presentations would mitigate the effect of earthquakes on the objects kept inside them.

In a museum, in addition to the safety of the staff and the visitors, the safety of museum objects is paramount. Normally, a restorer is trained to deal with expected, everyday situations. But damage due to an earthquake creates a special situation, requiring a different type of training. In fact, we faced this situation while trying to rescue damaged objects housed in Aina Mahal at Bhuj in Gujarat. Without the proper training, team workers cannot be expected to launch an effective response. There is an urgent need for training in response planning and in implementation of post-earthquake salvage and stabilization efforts. Conservators and restorers need to be familiar with the forms of recovery that can be undertaken during the first few hours and days following an earthquake, as well as during the subsequent salvage operations, which may last many months or even years. This project will involve complex training to help professionals face a very large range of challenges, since earthquakes are sometimes followed by fire, flood, mud slides, and security concerns. In the difficult times following a disaster, the shelter that the museum structure once provided is often no longer effective. Rain or dramatic environmental shifts threaten the safety of the collections.

Collections are often housed in historical buildings such as palaces, temples, and churches. There is a need to strengthen these buildings in order to make them earthquake resistant. Techniques are now available to improve the structural strength of such buildings. It is necessary to make this engineering knowledge as widespread as possible. When a new building is constructed, Indian seismic codes for making a building earthquake resistant must be applied. This practice should be especially followed for museums. Directors and authorities of museums should be made familiar with the basics of these codes so that they understand their value. Only when they understand, will they be in a position to support the higher cost of building or retrofitting a building meant to house cultural heritage, and in turn, they will be in a position to convince their trustees and governments that a responsible course of action has been taken.

Conclusion

It can be seen that by observing certain precautions, damage to museums and museum collections can be minimized and perhaps even prevented altogether. In summary, my recommendations are:

- Studies on earthquake safety of museums should be undertaken.
- A seismic design code for museum structures should be developed.
- Seismic zone maps should be regularly consulted before a new building is designed or an existing one is renovated.
- Training is required at several levels: for architects and engineers, for curators and directors, and for restorers and conservators.

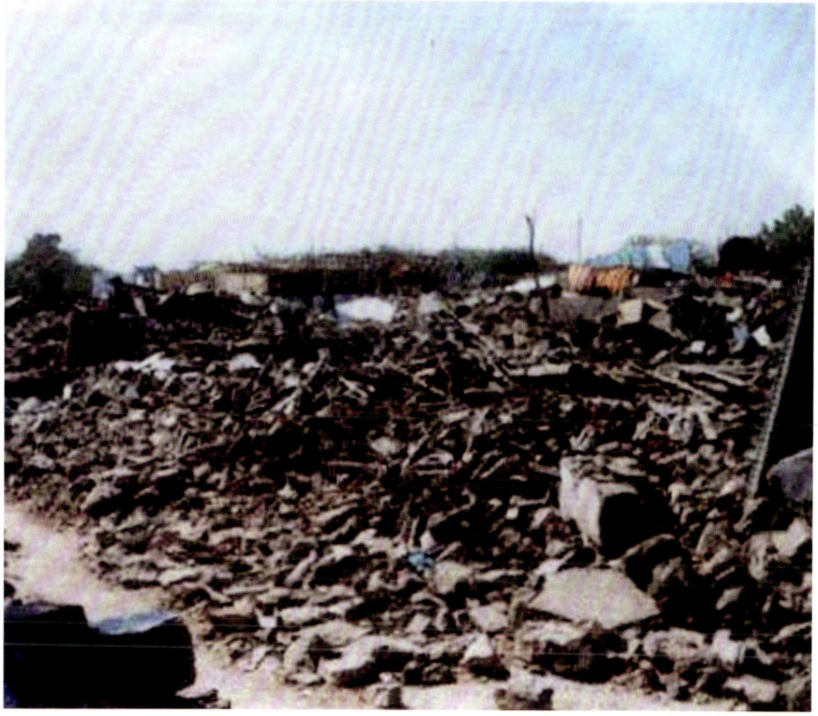

Figure 1
Damage in the village of Dhamadka.

Appendix 1: Earthquakes in India

June 16, 1819	Cutch	December 10, 1967	Koyna
June 12, 1897	Assam	March 23, 1970	Bharuch
February 8, 1900	Coimbatore	August 21, 1988	Bihar-Nepal
April 4, 1905	Kangra	October 20, 1991	Uttarkashi
January 15, 1934	Bihar-Nepal	May 22, 1997	Jabalpur
May 31, 1935	Quetta	March 29, 1999	Chamoli
August 15, 1950	Assam	January 26, 2001	Gujarat
July 21, 1956	Anjar	October 8, 2005	Kashmir

Appendix 2: Indian Seismic Codes

IS 1893 (pt. 1), 2002, *Indian Standard Criteria for Earthquake Resistant Design of Structures* (5th revision).

IS 4326, 1993, *Indian Standard Code of Practice for Earthquake Resistant Design and Construction of Buildings* (2nd revision).

IS 13827, 1993, *Indian Standard Guidelines for Improving Earthquake Resistant Earthen Buildings.*

IS 13828, 1993, *Indian Standard Guidelines for Improving Earthquake Resistance of Low Strength Masonry Buildings.*

IS 13920, 1993, *Indian Standard Code of Practice for Ductile Detailing of Reinforced Concrete Structures Subjected to Seismic Forces.*

IS 13935, 1993, *Indian Standard Guidelines for Repair and Seismic Strengthening of Buildings.*

Bibliography

EERI. *Earthquake Rebuilding in Gujarat, India,* an EERI [Earthquake Engineering Research Institute] Recovery Reconnaissance Report (Kanpur, National Information Center of Earthquake Engineering, Indian Institute of Technology, 2005).

Indian National Trust for Art and Cultural Heritage [INTACH]. "INTACH Assessment Report—An INTACH Initiative for the Conservation of Cultural Heritage in Earthquake Affected Gujarat," INTACH, New Delhi, 2001.

———. "INTACH Field Assessment Report—INTACH's Initiative: Study of the Impact of the Earthquake on the Cultural Properties in Jammu, Poonch and Riazi, J & K [Jammu and Kashmir]," INTACH, New Delhi, 2005.

McGuire, Robin K. *Seismic Hazard and Risk Analysis* (Oakland, Calif., Earthquake Engineering Research Institute, 2004).

Murty, C. V. R. *Earthquake Tips: Learning Earthquake Design and Construction* (Kanpur, National Information Center of Earthquake Engineering, Indian Institute of Technology, 2005).

National Information Centre of Earthquake Engineering [Kanpur]. *Guidelines for Earthquake Resistant Non-Engineered Construction,* rev. ed. of *Basic Concept of Seismic Codes,* vol. 1, pt. 2 (Tokyo, International Association for Earthquake Engineering. 1980).

Yashinsky, Mark, and M. J. Karshenas. *Fundamentals of Seismic Protection for Bridges* (Oakland, Calif., Earthquake Engineering Research Institute, 2003).

Yelland, Richard, and Brian Tucker, eds. *Keeping Schools Safe in Earthquakes,* 1st Indian ed. (Paris, Organization of Economic Cooperation and Development, 2006), pp. 229–35.

Seismic Mount Making: A Review of the Protection of Objects in the J. Paul Getty Museum from Earthquake Damage

McKenzie Lowry **David Armendariz**

B. J. Farrar **Jerry Podany**

Introduction

Providing suitable mounts for exhibited objects is one of the most important steps that can be taken for preservation of collections in seismically active areas. A well-designed seismic mount can mean the difference between an object that survives a devastating earthquake and one that is severely damaged or even totally lost. Such mounts need not be complex or costly to be effective, nor do they need to take a great deal of time to manufacture or install. To be effective, however, appropriate information regarding the site and the object itself must be taken into account as the mounts are conceived, built, and attached. This information should include the basic structural and seismic response characteristics of the building, the gallery, and the display furniture, as well as the object's material characteristics and overall mass distribution. Consideration should be given to the basic laws of physics. And a good measure of common sense is always useful.

In 1983 the J. Paul Getty Museum began its efforts to identify measures the staff could take to mitigate the effects of earthquakes and to protect its collections by addressing the following questions:

- Based on the overall geology and seismic history of the Museum site, what are the characteristics of the worst-case seismic event that might occur? These characteristics should be determined within a reasonable risk level and time frame.
- What would the overall response of the building be to such a worst-case event, and would any specific part of the building respond differently from any other part?
- How would the contents of the building (the collections, display furniture, and fixtures) respond to the earthquake motions and forces?
- How could any of these effects and responses be minimized to prevent damage to the collections?

The last two questions drove the efforts of the Getty Museum to develop and implement effective solutions to the problem of protecting the collections from earthquake damage. Those efforts, now some two decades old, required the input of many individuals and brought together a wide variety of necessary expertise. Curators and exhibition designers raised aesthetic and presentation concerns regarding the installations. Security implications for the collection were reviewed, as was the safety of the visiting public. The accessibility and visibility of the collection for educational needs were taken into account, and any impact on accessibility brought about by approaches to seismic mitigation was reviewed. Consulting engineers and seismologists brought much-needed expertise in the area of earthquake engineering and research. And concerns regarding the safety and long-term preservation of the objects remained, of course, paramount. All of these aspects came together in a complex exhibition and storage planning process that was made efficient and productive by the willingness of all participants to be flexible and to consider compromise, while keeping the ultimate goal of preservation in mind.

Seismic Protection

The seismic protection of the objects in the antiquities collection of the Getty Museum has been pursued through three approaches:

- Lowering the object's or object assembly's center of gravity, either by adjusting the proportions of the base-to-height (aspect) ratio or by adding weight. Examples of this strategy include placing sandbags inside vases, attaching objects to heavy bases, and adding lead bricks or other significant weights to the inside bottom of pedestals or cases.
- Restraining objects by firmly securing them to the floor, pedestals, shelves, wall, and/or supporting mounts.
- Assuring that the ground (building, floor, pedestal, or case) and the object will respond to earthquake forces (be moved) relatively independently of one another, and that, while the ground may move during an earthquake, these motions will not be fully transferred to the object or object assembly.

In the first of these approaches, the strategy of lowering the center of gravity by adding weight to an object or object assembly is fairly straightforward. In some situations, however, adding sufficient weight can become impractical and can sometimes threaten damage if the object is not structurally robust enough to accommodate the additional weight. In this approach, determining the amount of weight that is needed to achieve stability and prevent rocking or overturning during an earthquake is essential to an effective mitigation effort.

The second approach, anchoring the object, requires that the object have sufficiently high inherent strength and rigidity (or have these qualities enhanced by a supporting mount) to withstand the transferred earthquake forces without risk of damage. Such mounts may also limit the motion of the object and minimize the effects of dynamic forces.

The last of the three approaches—isolating the object from earthquake forces (transferred through the building) by allowing the building and the object or object assembly to move somewhat independently—is normally achieved by a base isolator or Teflon pads to encourage sliding. This approach should only be undertaken when there is significant knowledge of the characteristics of the expected earthquake and building response. Sufficient safety factors must be included, so that the forces of the earthquake are statistically unlikely to exceed those that the isolator was designed for.

Defining the Seismic Threat

None of the approaches discussed above should be undertaken without at least a rudimentary understanding of the threat. That is to say, one should have a general understanding of how earthquakes affect structures and contents. It is equally necessary to seek out the assistance of experienced seismic engineers and seismologists in order to establish the characteristics of the worst-case scenario earthquake predicted for the area and specific site where the collection under consideration is housed. Understanding the seismic threat is the first step toward mitigating its effects.

The duration, strength, frequency content, and potential for displacement of the design earthquake—as well as the response of the building and, indeed, the specific areas of the building where the collections are housed—will give direction to any efforts in developing mitigation approaches. It is by far more important to know the peak acceleration, velocity, and predominant period of an earthquake (and the building and object response) than it is to know the expected Richter scale magnitude, for example. Insufficient design could be useless during an event—or even make things worse. It should also be noted that overdesign, though it is the common response to threats in the museum world, could be both dangerous to the object and a waste of resources. As a result, well-developed parameters are a necessity and a matter of practicality.

While accurately predicting the specific nature of any future earthquake remains as elusive as predicting the exact time an earthquake will occur, geological records, seismic history, identification of active or potentially active faults, and a full structural study of the building and its potential response can provide the appropriate parameters for mitigation design. In most cases the study of these factors leads to a statistical estimate and characterization of the worst-case scenario in a range of years and probability. Commonly, a 50%, 20%, or 10% probability in a 50–100 year period is sought. As the percentage of likelihood goes down and the time period lengthens, the strength and presumed destructive power of the earthquake normally rise. An evaluation of what commitment can be made to seismic mitigation efforts, balanced with what risks those responsible for a collection are willing to accept, must be considered. Museums, dedicated to preserving the cultural heritage in their care, are commonly averse to risk and to loss at any level—though they might not always be fully realistic.

Earthquake Damage Mitigation at the Getty Museum

In an effort to define the seismic threat at its location, the J. Paul Getty Museum at the Villa, in Malibu, California, commissioned Lindvall, Richter and Associates in 1984 to prepare a geologic and seismologic study of the Museum site and a geotechnical and structural response study of the Museum building (Lindvall, Richter and Associates 1984). The Museum established an acceptable risk level, defined as an event with an 80% probability of not being exceeded in a 50 year period. Reoccurrence of such an event was estimated to be every 225 years. The study identified two events that would have the most impact upon the Museum: (1) an 8.3 Richter scale earthquake on the San Andreas Fault, some 67.5 km (42 mi.) away from the Museum, that would result in a horizontal ground acceleration of 0.2 g at the Museum site, and

(2) a 6.5 Richter scale earthquake on the Malibu Coast–Santa Monica fault system at a distance of 1.6 km (1 mi.) from the Museum that would potentially produce a maximum 0.7 g horizontal ground acceleration at the Museum site.

Since the main Museum building proved to be an impressively stiff structure, with a natural period of 0.1 sec., it was assumed that the ground accelerations would be directly transmitted to the structure with little damping, and hence be transferred to the objects in the collection. Because of this possibility, the collections at the Museum were considered at great risk unless some sort of mitigation effort were implemented.

The conclusions of the Lindvall, Richter report provided dynamic data for use in analyzing the behavior of art objects housed at the Museum. The study had great importance for the future plans of the Museum and was the basis for future mitigation studies, such as a report by Agbabian, Masri, and Nigbor (1990) that attempted to predict the seismic response of artworks by modeling generic categories that represented whole groups of objects. The study divided these groups into two main overall categories of rigid and flexible. However, the complexities presented by modeling nonlinear responses, as well as the many unknowns of the material properties and strength characteristics of the objects themselves, limited the application of the study. Additionally, the calculated responses of objects have not always been borne out in tests. It is difficult and often impractical to model and then actually test every object or object assembly. Evaluations can also be complex because of the large number of influencing variables for each unique object. For example, the earthquake response of a rigid block (the object or object assembly) depends upon a number of things:

- duration and nature of the motion of the earthquake forces
- geometry of the object or object assembly (aspect ratio)
- size of the object or object assembly
- properties of the contact surfaces (coefficient of friction)
- material properties of an object (what it is made of and/or how it has altered over time)

Nonetheless, a limited number of basic assumptions and general guidelines applied to appropriate stability measures are enormously useful. From the Lindvall, Richter report, as well as the Agbabian, Masri, and Nigbor study, a basic criterion for stability has evolved.

For example, the response of a rigid object to earthquake-induced forces and motion can be sliding or, if the friction between the object and the supporting plain is high enough, rocking and eventual overturning. Rocking and overturning are based both on the nature of the earthquake acceleration and the object's (or object assembly's) geometry and mass distribution. Figure 1 shows a rocking stability chart based on the data from the design earthquake for the Getty Villa; it applies to objects subjected to an event based on that data.

Rocking, sliding, or overturning can be predicted for objects of simple shape by determining the base-to-height ratio. Rocking and sliding will occur when the maximum horizontal acceleration (a_{max}) is greater than the base-to-height aspect ratio (the ratio of the base width [B] to the height of the object assembly [H] in centimeters): $a_{max} > B/H$. For the Villa site, the primary horizontal component peak ground acceleration is assumed to be 0.7 g. Overturning

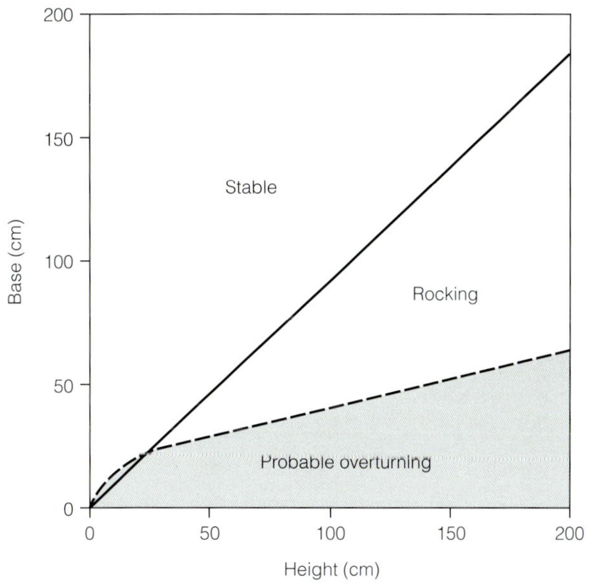

Figure 1
Rocking stability chart based on data from the representative (design) earthquake for the Getty Museum at the Villa. Primary horizontal component peak acceleration is 687 cm/sec^2 (22.5 ft./sec.^{2}), and peak velocity is 33.9 cm/sec. (1.1 ft./sec.). Reprinted from Agbabian, Masri, and Nigbor (1990, 38).

will occur when the relationship of the aspect ratio to the velocity of the earthquake (V_0) can be described as follows: $V_0 > 10B/\sqrt{H}$ (the peak ground velocity for the Villa site is assumed to be 33.9 cm/sec., or 1.1 ft./sec.), when dimensions are given in centimeters. Whether rocking, sliding, or overturning occurs depends upon the *B*/*H* ratio and the location of the center of gravity, as shown in figure 2. Even if the aspect ratio is favorable, sliding is predominantly dependent upon the coefficient of friction between the bottom of the object and the surface it rests upon. If the two surfaces are smooth, sliding is likely; if they are rough or resistant, rocking is more likely.

While it should be kept in mind that rocking, sliding, and even overturning are nonlinear and complex responses, basic rules of thumb remain important in mitigation efforts, and it is important to learn the peak velocity

Figure 2
Objects of different aspect ratios and their probable response to earthquake forces. Whether an object will overturn or slide is directly related to the aspect ratio. The object on the left will most probably rock and/or overturn, while the object on the right, with a lower center of gravity and a more favorable aspect ratio, is more likely to slide, assuming that the coefficient of friction between the object and the support plane is sufficiently low (cg = center of gravity).

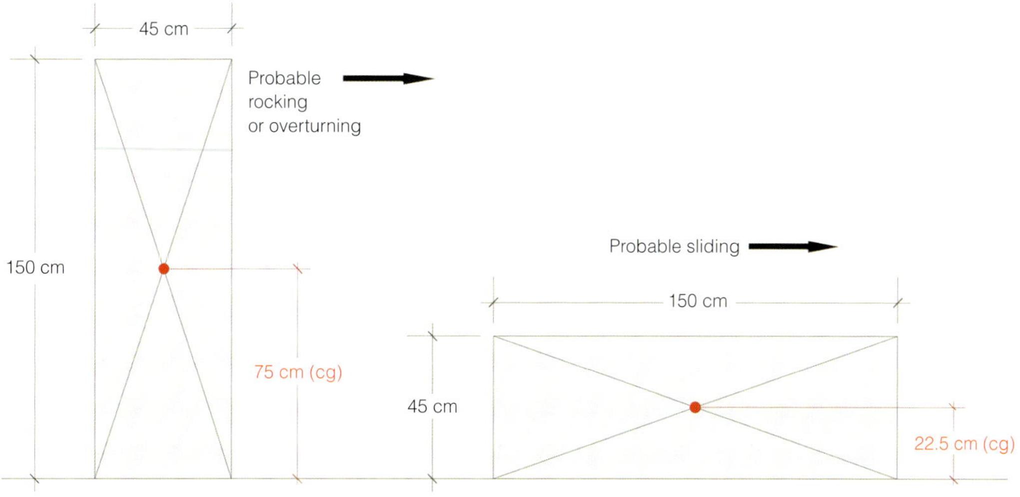

and acceleration of an earthquake—two of the most useful items of data related to the potential threat.

It also should be noted that only rigid systems are addressed in this paper. If an object is not rigid, then every effort should be made to introduce a mount to provide rigidity and thus greater protection and predictability of response.

Determining Stability through the Effective Aspect Ratio

Complex objects or object assemblies can be represented as simplified volumetric models known as equivalent blocks. These models are geometric representations of the distribution of mass and the centers of gravity of the actual objects or object assemblies. The effective aspect ratio for an equivalent block is arrived at through the following formula:

$$0.5\ B/H_{cg}$$

The initial studies at the Getty Museum at the Villa targeted an effective aspect ratio of 2 for freestanding objects and object assemblies. This guideline was arrived at by multiplying a preferred effective aspect ratio of 0.7 (equal to the peak ground acceleration of 0.7 g) by a safety factor of 3. Although these parameters would ensure adequate stability in a large earthquake, it is obviously pragmatically very difficult to achieve within the confines of museum display. Inherent proportions of an object are not alterable, since they are part of the object's unique aesthetic and historical presence. At best, the object might be rigidly attached to a supporting base or plinth in order to arrive at the aspect ratio and lower the center of gravity of what is now an assembly. But even in such a case, aesthetic and practical limitations influence the possibilities and outcome. A heavy marble torso, for example, might require an extremely wide or short pedestal, resulting in a rather unattractive presentation that may also require an unacceptably large amount of gallery space.

Altering the aspect ratio and/or adding weight to the lower sections of the assembly or object are two ways the center or gravity can be altered (lowered) to achieve greater stability. Before this is attempted, however, calculating the location of the center of gravity of any object or assembly is essential.

Determining the Center of Gravity

The term *center of gravity* (cg) describes a theoretical point within the mass of an object or object assembly where all earthquake forces are focused. It is, in some ways, the true center of mass of the object or assembly. The lower the center of gravity, the more stable and resistant to rocking and overturning the object or assembly will be. Determining the location of the center of gravity can be achieved in a number of ways. One of the simplest methods is to measure the maximum depth, width, and height of the object and then translate those dimensions into a geometric volume that closely resembles the object's shape, assuming an even weight distribution throughout the volume of the object. The center of gravity will be roughly the same as the calculated center of the geometric volume (fig. 3).

When an object is composed of segments with differing dimensions or densities, each segment should be translated into a geometric shape and the

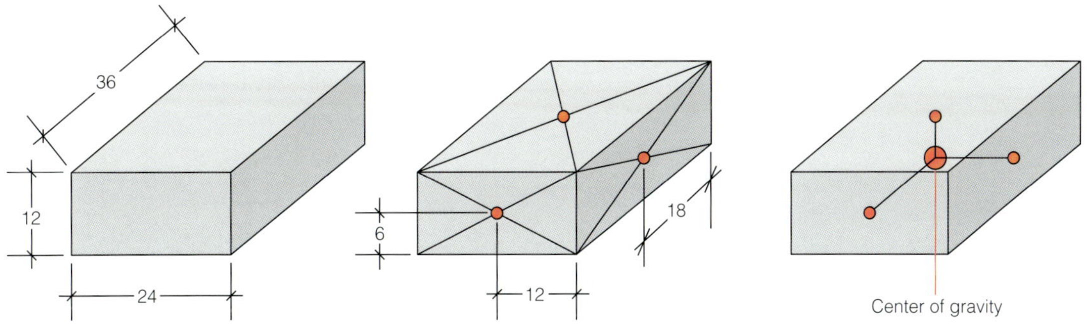

equivalent block determined (Podany and Leavengood 1992). In these calculations the center of gravity is arrived at by taking into consideration the weight of the object or weights of the various segments of the object or object assembly (fig. 4). These methods can be advantageous when an object is complex in shape, is made of a variety of segments, has an eccentric distribution of mass, or is part of an assembly (such as a sculpture and pedestal combination). If the separate components of an object cannot be weighed, calculations can be made based on references of material properties (marble, for example, generally weighs approximately 2306.6 kg/m^3, or 144 lb./ft^3).

The following is an example of the method for determining the height of the equivalent block (H_{cg}) and center of gravity (in the examples that follow, measurements are given in metric units):

$$H_{cg} = \frac{(D1 \times W1) + (D2 \times W2)}{W1 + W2}$$

where D = distance from display surface to each section's cg and W = each section's weight.

If: $D1$ = 44.5 cm and $W1$ = 18 kg, and $D2$ = 16.5 cm and $W2$ = 66 kg, then

$$\frac{(44.5 \times 18) + (16.5 \times 66)}{18 + 66} \quad \text{or} \quad \frac{1890}{84} = 22.5$$

Figure 3
A simplified method for determining the center of gravity, which involves finding the center of a geometric volume. This procedure assumes, however, that the simplified geometric model accurately reflects the mass of the object and that the density of the material is consistent throughout.

Figure 4
A method for determining the equivalent block of a complex sculpture or a sculpture with varying densities and eccentric parts (W = weight). *Bust of an African.* Roman, second half of second century A.D. J. Paul Getty Museum, 71.AA.437.

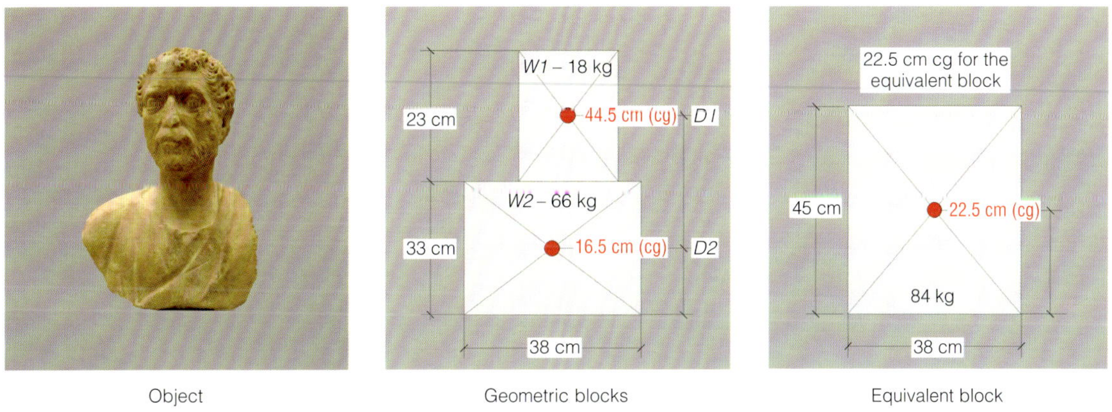

Thus, the equivalent block has a center of gravity at 22.5 cm, and $2H_{cg}$ gives us an equivalent block with a height of 45 cm.

From this point, the *effective aspect ratio* of the equivalent block can be determined by the following formula, which divides one-half of the base dimension of the equivalent block by the height from the ground to the center of gravity of the equivalent block:

$$\frac{0.5B}{H_{cg}}$$

Determining the Effective Aspect Ratio

Using the effective aspect ratio formula, $0.5B/H_{cg}$, we find that the unit has an effective aspect ratio of 0.84:

$$\frac{38}{2} = 19 \qquad \frac{19}{22.5} = 0.84$$

An additional example is illustrated in figure 5. This example uses the weight and cg of an object and the weight and cg of a pedestal to find the cg of the assembly and the equivalent aspect ratio.

$$H_{cg} = \frac{(D1 \times W1) + (D2 \times W2) + (D3 + W3)}{W1 + W2 + W3}$$

where D = distance from the ground to each section's cg and W = each section's weight.

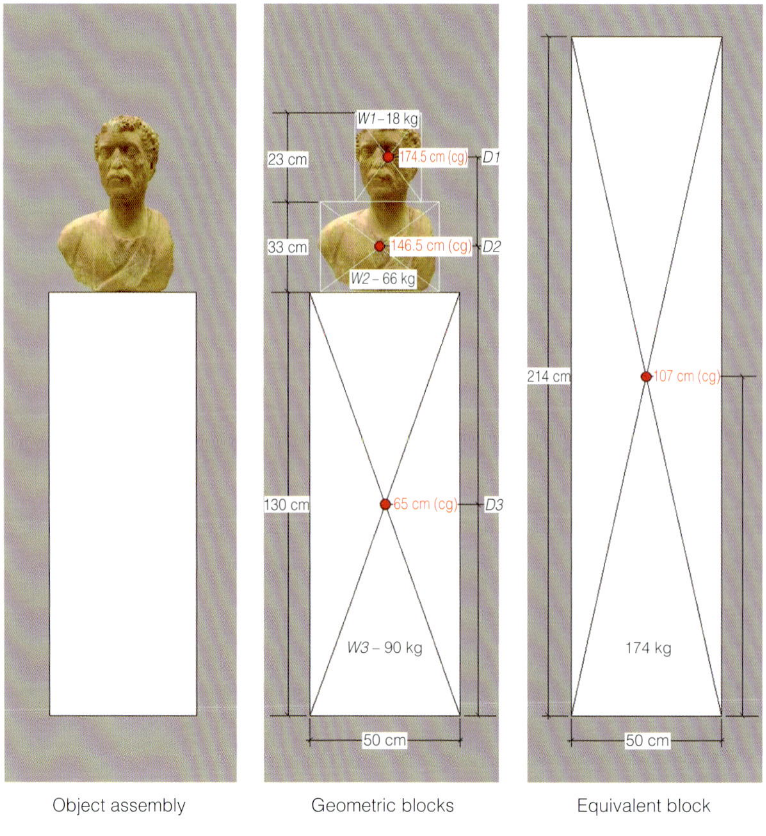

Figure 5
A method for determining the equivalent block of an object assembly—in this case, a sculpture and pedestal combination (W = weight).

If: $D1$ = 174.5 cm and $W1$ = 18 kg, $D2$ = 146.5 cm and $W2$ = 66 kg, and $D3$ = 65 cm and $W3$ = 90 kg, then

$$\frac{(174.5 \times 18) + (146.5 \times 66) + (65 \times 90)}{18 + 66 + 90}$$

or $$\frac{18{,}660}{174} = 107.2$$

Thus, the equivalent block has a center of gravity at 107 cm, and the $2H_{cg}$ formula gives us an equivalent block height of 214 cm.

Using the effective aspect ratio formula, $0.5B/H_{cg}$, we find that the unit has an effective aspect ratio of 0.23:

$$\frac{50}{2} = 25 \qquad \frac{25}{107} - 0.23$$

Lowering the Center of Gravity by Adjusting the Proportions of the Display Furniture

If a center of gravity is found to be too high to achieve the desired degree of stability (based on the type of data listed in fig. 1 and developed for the specific site and design event), the center of gravity can be lowered by a number of approaches. For example, if an object is attached to a base or pedestal, there are three fairly simple methods for lowering its center of gravity. The base or pedestal can be made wider, or it can be weighted or secured to the floor. If circumstances allow, fastening to the floor provides the greatest stability of any option. However, this approach assumes that the object is sufficiently structurally sound and can withstand any earthquake forces transmitted through the building, pedestal, and mount.

If anchoring to the floor is not an option, the object can be attached to a base or pedestal of appropriate width to achieve the desired aspect ratio. Taking the previous example, which is illustrated in figure 5, one can see that the assembly has an effective aspect ratio of only 0.23, using the $0.5B/H_{cg}$ formula described earlier.

To achieve an aspect ratio of at least 2, the base of the equivalent block must be at least double its height:

$$214 \text{ cm} \times 2 = 428 \text{ cm}$$

While the resultant pedestal dimensions of 2 m (6 ft., 6 in.) in height and 4 m (13 ft.) in width and depth may be stable, such a pedestal would clearly not be very practical.

An alternative is to attach an appropriately sized baseplate or an enlarged base to the bottom of the pedestal. While this method would reduce the overall volume of the pedestal, it would also remain impractical for most exhibition environments, and it is unlikely that the baseplate dimensions necessary for full stability would be aesthetically appropriate.

There are instances, however, when an enlarged plinth becomes an effective option for stabilizing a group of smaller pedestals. As shown in figure 6, the center of gravity remains the same, even though there are two pedestals attached to a single base. The narrowest dimension of both pedestals remains

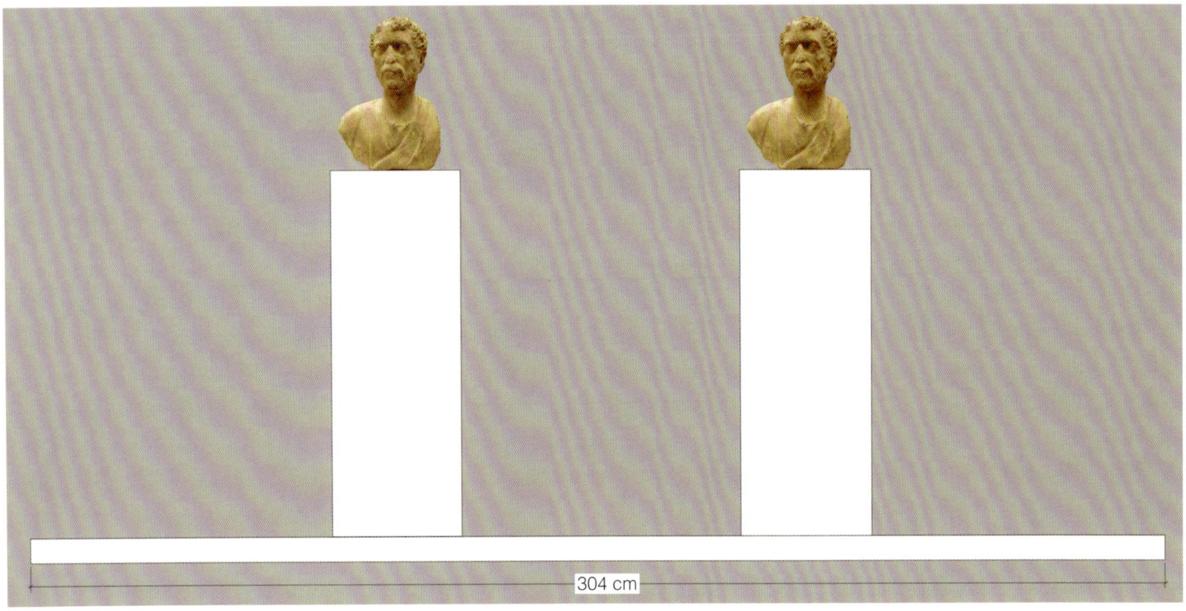

Figure 6
A wide base structure that stabilizes multiple object assemblies anchored to the base. The individual object assemblies have less favorable *B/H* ratios.

50 cm (19 11/16 in.), and the height remains unchanged. Great care should be taken to understand the relationship among multiple elements when they are combined. The basic concept is to keep the cg as low as possible; therefore, the taller a display is, the wider the footprint of the base will need to become.

Lowering the Center of Gravity by Adding Weight to the Assembly

The center of gravity can also be lowered by distributing the majority of the weight toward the bottom of the assembly. Adding weight to the base or lower part of the pedestal, therefore, provides a third option. It should be noted, however, that the added weight, or ballast, ultimately takes up space and changes the mass distribution of the form in an otherwise limited width-and-depth profile of the pedestal or case construction. There is a limit to how high the profile of the added weight can become before it defeats the purpose of its addition. When calculating the center of gravity and the equivalent block of an assembly to which weight has been added, the weight mass should be seen as a separate dimensional element, and the height of the assembly (pedestal or case structure) should be reduced accordingly in its geometric representation. It is usually more effective to increase the footprint of the pedestal or support structure while also adding ballast. Referring again to figure 5 (effective aspect ratio less than 2), the amount of additional weight needed in the base to lower the cg can be calculated as follows (a number of lead bricks will be used in this example for the added weight). In this example, illustrated in figure 7, all of the components are rigidly and securely attached together.

$$H_{cg} = \frac{(D1 \times W1) + (D2 \times W2) + (D3 \times W3)}{W1 + W2 + W3}$$

$D1$ = 152.5 cm (cg of object-equivalent block [fig. 4] plus the height of the pedestal assembly [pedestal and plinth])

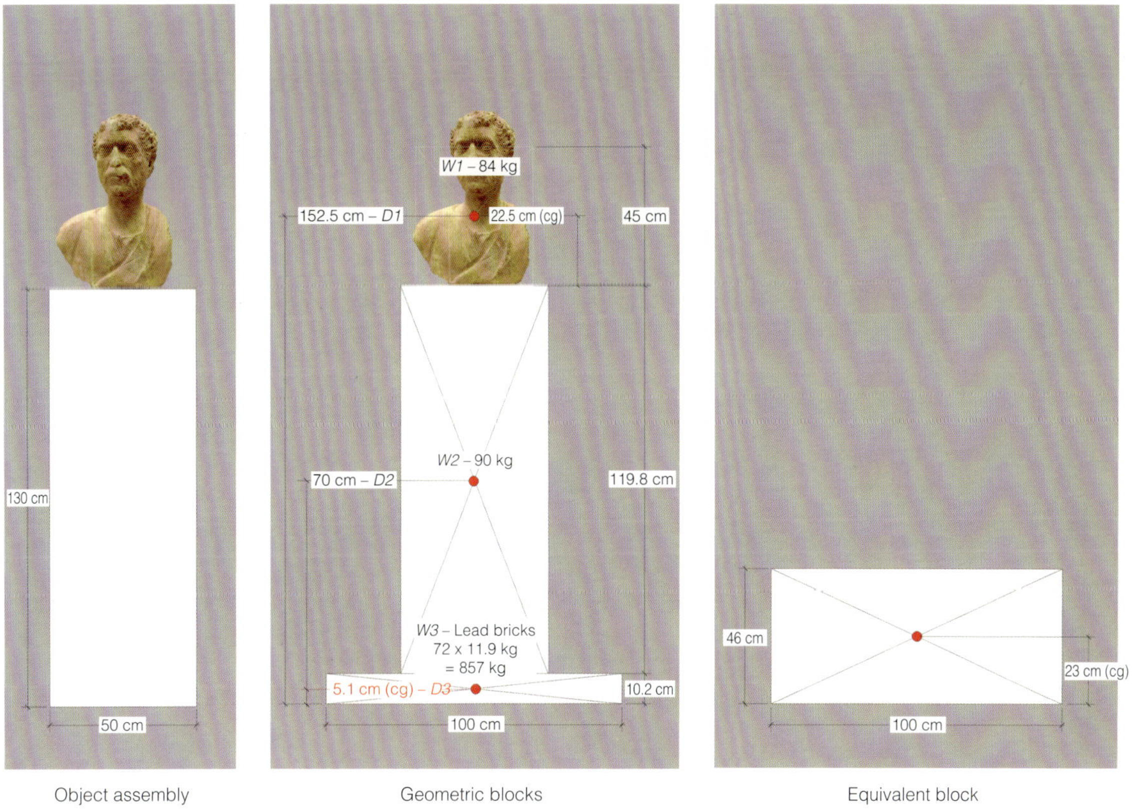

Figure 7
Increase in stability through alteration of the equivalent block. In the system's new equivalent block, the cg has been lowered to 23 cm (9 in.) by adding 857 kg (1885.4 lb.) and increasing the base by 50 cm (19 11/16 in.). Note that the center of gravity must be calculated for the added weight mass—in this case, lead. Stacking the weight too high will cause the mass to become unstable.

$D2 = 70$ cm (cg of pedestal [height of pedestal minus the height of the added weight or plinth, divided by 2 plus the height of the added weight or plinth])

$D3 = 5.1$ cm (cg of plinth/added weight equivalent block)
$W1 = 84$ kg (weight of head)
$W2 = 90$ kg (weight of pedestal)
$W3 = 857$ kg (weight of plinth/added weight assembly)

Then, $$\frac{(152.5 \times 84) + (70 \times 90) + (5.1 \times 857)}{84 + 90 + 857}$$

or $$\frac{23{,}481}{1031} = 22.8$$

Recalculating the effective aspect ratio shows that it is now over 2:

$$\frac{100}{2} = 50 \qquad \frac{50}{23} = 2.1$$

Sufficient room inside the pedestal or plinth must be available to accommodate the added weight (ballast) that is meant to lower the center of gravity of the assembly. It should be noted that if the ballast does not completely fill the interior of the pedestal or plinth, the mass (dimensions and total weight) of the ballast and the mass of the plinth or pedestal must be treated as separate elements in the use of the equivalent block equation (as seen, for example, in fig. 7). It should also be noted that the ballast should be distributed in as low a profile as possible, since as the height of the ballast increases, the stability of the assembly decreases.

Although these examples have outlined the limitations of tall and narrow displays and have suggested methods of correcting this inherent instability (by adjusting the dimensions of the pedestal or pedestal base or by adding more weight), anchoring the object or the object assembly (including pedestals) to the floor or wall will achieve the same result. Anchoring may also minimize the need for added ballast in the case of pedestals and thus minimize the concentrated load on the floor area.

Seismic Mount Making

The previous discussion assumes that the object is sufficiently robust and rigid to withstand any transmitted earthquake force. But this is often not the case. Additional strength can be provided, however, by introducing supportive mounts that both cradle and restrain the object while making it possible to display the object in positions favorable to an effective display. The design and production of these fixtures, or mounts, in a wide array of complexity and materials—and within the context of seismic concerns—can be termed seismic mount making.

Any successful mount design requires a sufficient understanding of the dynamic forces imposed on the object and mount during an earthquake. But the inherent physical characteristics of the object are also critical. For example, the distribution of the object's volume and weight, as well as its material strength, must be kept in mind. The craft of mount making necessitates familiarity with diverse materials, including a wide range of metals, woods, plastics, synthetic composites, and fabrics. Mounts should always be made of strong, stable, and safe materials that do not adversely affect the objects or the environments within which the objects are exhibited. This precautionary characteristic is applicable to paints, elastomers, and other synthetic products as well. The materials must be nonabrasive, noncorrosive, stable, and nonstaining, and they must not exude corrosive vapors. Aesthetic considerations, which dictate how visually integrated the mount should be with the object itself, will also influence material choices.

In the design of a mount, the contact point between a mount and an object should be sufficiently large, as intimate as possible, and always nonabrasive to the object's surface. Small contact points result in higher forces that are concentrated at the points of contact. Conversely, the larger the contact point, the more the forces will be distributed and the more the mount-to-object connection will be secure.

As far as possible, mounts should be designed in such a manner that the object can be quickly, easily, and safely (without undo risk) removed from the mount. Adhering a mount to an object should be avoided whenever possible. If a mount must be adhered to an object, the adhesive should be stable, sufficiently strong, and easily reversible.

Mounts should never be weaker than the inherent strength of the object, and they should be sufficiently strong to resist the forces that will be imposed by the weight of the object combined with the forces of the earthquake. One would not, for example, support a marble bust with a material of lesser compressive, tensile, and shear strength than that of the marble itself. Generally, a safety factor of 3 is considered good practice, since dynamic forces due to earthquake motions can increase the total load on a mount by several magnitudes. This approach might translate, for example, into using a hook with an ultimate strength of 68 kg (150 lb.) to hang an object weighing 22.68 kg (50 lb.).

While strength is essential to the integrity of a mount, rigidity also plays a large role in the mount's ability to protect an object. The material used for a mount should be as rigid as possible, either because of its inherent strength or because it is manufactured with sufficiently thick stock to assure suitable stiffness.

Seismic mounts can be divided into two major categories: stationary/supportive and decoupling. The essential difference between these two categories is that a stationary/supportive mount fixes the object firmly in place, resulting in synchronous movement of the object and the mount with the pedestal, case, wall, floor, or building. In this case, the object is considered to be part of the building with respect to earthquake forces and motion. A decoupling mount allows the object to move within a defined parameter, as well as at a defined rate of speed, and with a predefined degree of ease or resistance dependent upon the amount of displacement and the characteristics of the design earthquake spectra being referenced. It is more accurate, however, to think of this situation not as a case of the object being allowed to move; instead, it can be understood as the building or ground moving beneath the object, without dramatically affecting the position of the object relative to its starting point before the earthquake. These decoupling devices are often referred to as isolators.

Stationary/Supportive Mounts

While stationary/supportive mounts restrain objects and minimize dynamic loads that might be caused by rocking, falling, or impact, it is very important that the mounted object be sufficiently robust in nature to withstand a considerable proportion of the seismic forces transferred through the building structure. Confidence in anchoring and supporting depends upon establishing or estimating the yield strength of the object, the anchor, and the mount (as well as what the object is anchored and mounted to). Confidence is also dependent upon the object being firmly held to the mount at suitable points and over a suitably large area, as well as dependent on the mount being securely fastened to the exhibition furniture, the wall, or the floor. The system must be rigid and respond as a single unit.

Although a number of smaller objects on display at the Getty Museum are restrained with very simple anchoring methods, such as wax or synthetic monofilament, more substantial mounts are the more common approach.

Interfaces Objects that do not sit in a stable and level position, that have uneven contact with the supporting plane (floor, pedestal top, or case deck), or that have the majority of their weight concentrated on small points (point loaded) require a custom-made interface to distribute the load evenly over the bottom surface or over the surface on which the object rests (fig. 8). Such interfaces are made of materials with relatively high compression strength, such as filled epoxies. Prior to the casting of an interface, the underside of the

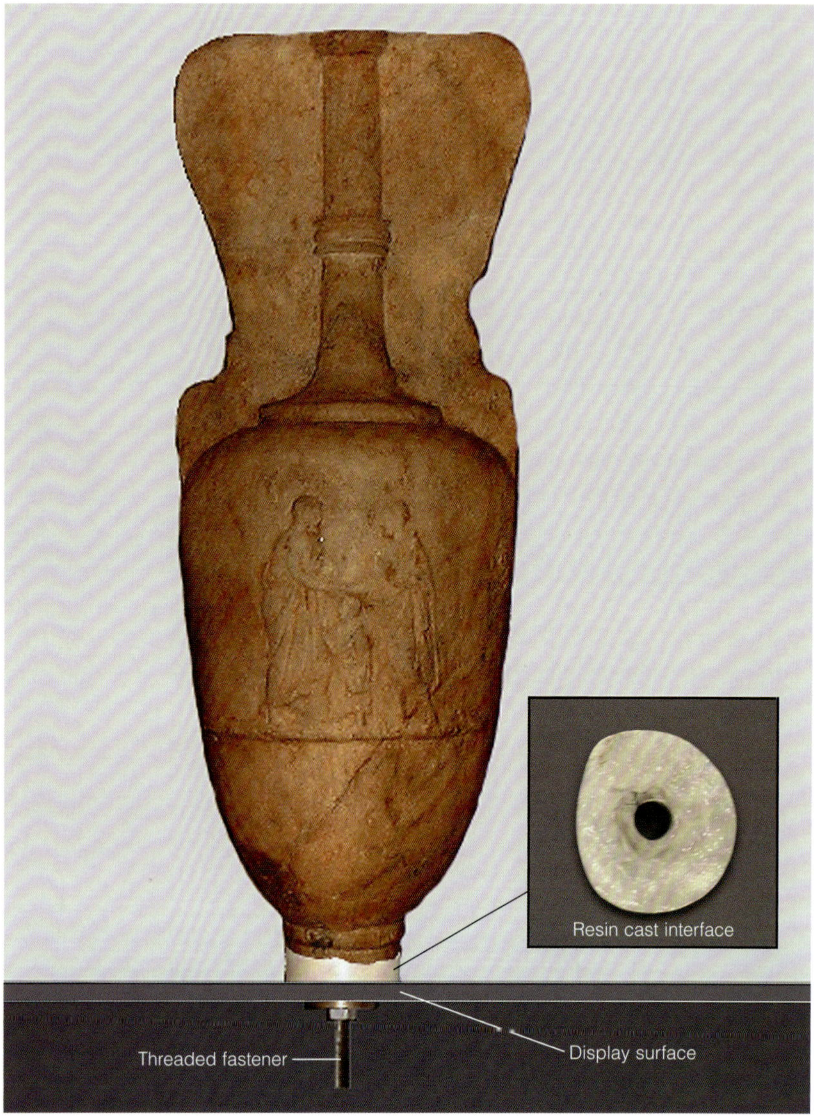

Figure 8
An interface supporting the broken bottom edge of an object. *Funerary Loutrophoros of Aristomachos*. Greek (Attic), ca. 370 B.C. J. Paul Getty Museum, 83.AA.253.

object's base should be inspected for undercuts or cavities to avoid any physical locking of the interface to the piece. Caution should be taken as well when employing fast curing epoxies, since they generate substantial amounts of heat during curing.

To cast a typical interface, a modest amount of thixotropic epoxy (enough to cover the bottom of the object and fill any voids) is placed on a non-stick surface and covered with a sufficiently thick barrier of thin plastic film (such as the type used in the food service industry). The object is then lowered onto the plastic film and allowed to settle until the desired orientation is achieved and the excess epoxy is displaced. At this point the object should be secured so that it does not shift while the epoxy is curing. Care should be taken to assure that none of the epoxy has come into direct contact with the object. The epoxy should also be monitored to prevent it from slumping away from its intended supportive shape. Once the epoxy is fully cured, the object is lifted away, the plastic wrap removed, and the interface trimmed to the desired shape. In most cases the interface will be secured to the display deck rather than to the object.

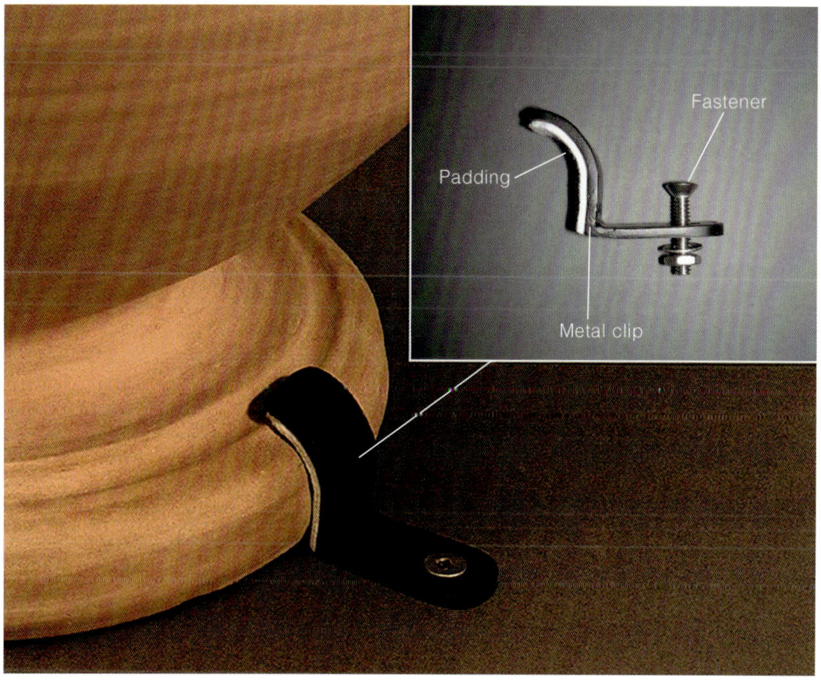

Figure 9
A typical clip construction.

Clips Clips are relatively small point-of-contact mounts that restrain an object's movement but do not provide any support (with the exception of wall-hung objects). Normally, three or more clips are used for any single object being restrained (fig. 9 and figs. 10a and 10b). For objects that are mounted to the wall, the clips may be equated to frame clips, which support the object vertically (along the lower edge) as well as lightly compress the object against the wall. In the instance of freestanding objects, clips are normally applied to the object's base or lower edge. When using clips in such a manner, caution must be exercised, since there will be considerable stress at the point where the object is anchored during an earthquake. Such stresses are due to the fact that the object acts as a kind of lever with a pivot point at the anchored, clipped location. The taller the object is (that is also to say, the higher the cg), the greater the forces will be at the anchor (clip) points. Significant accelerations and the repetitive reversal of direction occurring during an earthquake will cause considerable stress to build at the narrowest cross sections below the cg of the object. Stress failure at the area of high stress concentration (point loads caused by the clips) or at the area of material weakness is highly likely. When considering the use of clips, an evaluation of whether the object sits flat and evenly on its base is important. If the base is uneven, the use of clips around its perimeter

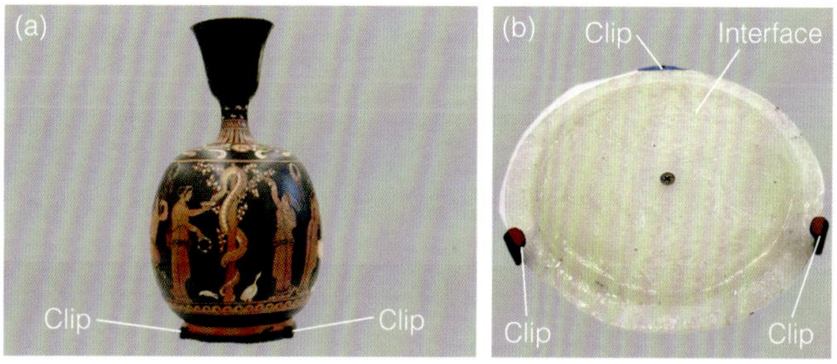

Figures 10a and 10b
An object secured with clips. A side view shows the location of the clips (a), and a top view of the interface without the object (b) shows the position of the three clips on the mounting deck. *Squat Lekythos*. Greek (Paestan), 350–340 B.C. Lent by the Republic of Italy.

can lead to severe damage, with the possibility of two opposing clips pulling down on unsupported edges. Casting an appropriate interface, as described above, is necessary in such a case (figs. 10a and 10b).

A three-point restraint is recommended when using clips to secure objects with round or oval bases. In these cases the clips are spaced as close to 120° apart from one another as possible. For square or rectangular bases, a four-point restraint is recommended, with clips on all four sides or at each corner. Each clip should be fabricated in two parts, a contour clip form and a base (fig. 9). The contour should be bent to follow the profile of the base, though it should be slightly larger to allow the clip to slip onto the object from the side. Once the contours are finished, they should be either soldered or welded to a base piece that has drilled and countersunk holes for anchoring screws.

The edges of each clip should be ground down and slightly rounded, and an appropriate felt or padding should be applied to the interior faces to protect the object's surface.

Contour Mounts A contour mount is a supportive restraint that closely follows the exterior contour of the object, providing along its length either complete contact or points of connection to the object. A contour mount is made of material with the strength and stiffness necessary to support the object as well as withstand the earthquake load.

In most cases there are four basic materials used to fabricate contour mounts: steel (including stainless steel), brass, aluminum, and acrylic. Steel and brass are typically bent and/or welded to attain any given shape, whereas aluminum and acrylic parts are cut and/or adhered together to follow a predetermined profile.

When steel or brass are used, the object should be laid horizontally on a padded surface, with a strip of thin felt placed on the surface to be mimicked. A termination point should be identified near the top of the object where the contour will end. This location should coincide with a point where monofilament can be wrapped around the object and then tied to the contour mount as a passive restraint. A similar area should be identified near the bottom of the object.

A flexible cloth measuring tape can be used to determine the length of stock required to manufacture the mount. The stock can then be bent or cut to mimic the contour of the object. When the contour has a good fit and is in place, a straightedge can be held against the bottom of the piece and a mark made on the mount contour to indicate where it can be cut. The contour should then be

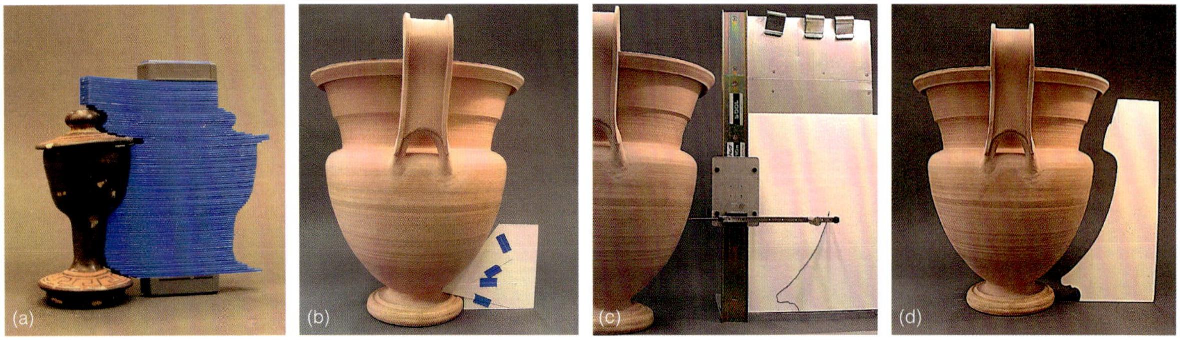

Figures 11a–d
Three methods for obtaining a profile: using a plastic profile gauge (a), using cut cardboard pieces (b), and using a custom-built profilometer (c). The resulting profile, cut in cardboard, is shown (d).

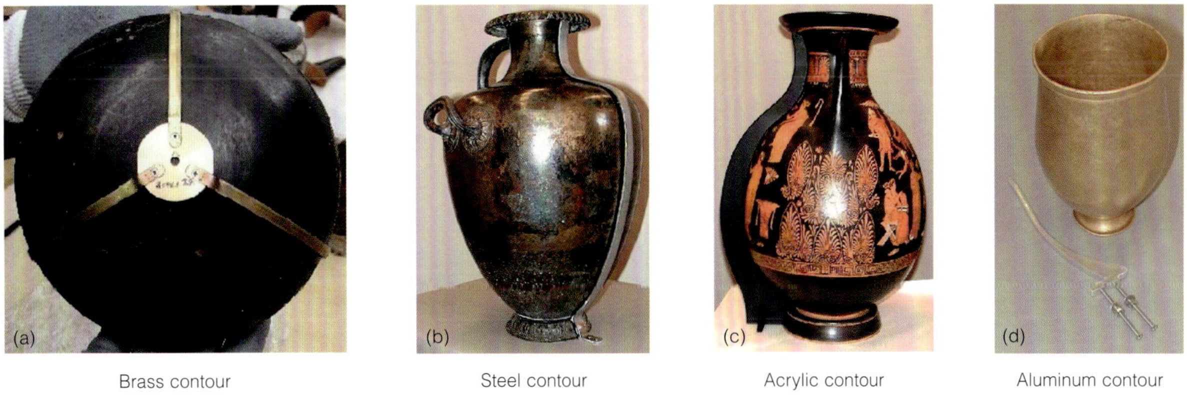

welded, soldered, adhered, or mechanically fixed to a predrilled base. When aluminum or acrylic are used, the contour shape must be cut from stock material rather than bent. To begin, a measurement of the object's profile is attained using a profilometer or a plastic contour gauge or by cutting out and piecing together sections of stiff paper or cardboard as shown in figures 11a–d.

Holes should be drilled through the mount at the previously determined locations near the top and bottom; monofilament is then inserted through these holes to secure the object to the mount. The interior surfaces should be felted, and exterior surfaces should be suitably finished (figs. 12a–d and 13).

Anchors If an object has existing holes from previous mounting or restoration efforts or if it is possible to safely drill appropriate holes to accommodate a mounting pin, it is advisable to anchor the pin into the object using

Figures 12a–d
Four different styles of contour mount made of different materials.
(a) *Phiale Mesomphalos (Six's Technique Phiale)*. Athenian, ca. 510–490 B.C. Antikensammlung, Staatliche Museen zu Berlin Preußischer Kulturbesitz.
(b) *Hydria Kalpis*. Greek, mid-fourth century B.C. J. Paul Getty Museum, 79.AC.119.
(c) *Apulian Red-Figure Pelike*. Greek (Apulian), ca. 340–330 B.C. Lent by the Republic of Italy. (d) *Beaker*. Roman, first century A.D. J. Paul Getty Museum, 2001.6.

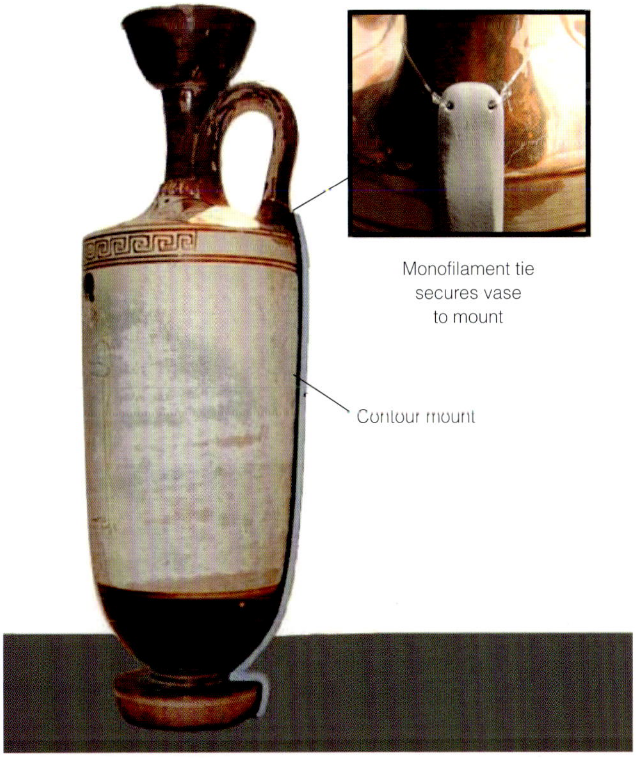

Figure 13
Monofilament tied to the object from the end of a contour mount. *White-ground Lekythos*. Greek (Attic), 460–450 B.C. J. Paul Getty Museum, 96.AE.99.

anchors into which the pin is threaded, rather than adhering the mounting pin directly into the object. Great care must be taken to thoroughly evaluate the fabric of the object surrounding the intended anchor points, ensuring that the material is sufficiently robust to withstand any seismic forces. Most off-the-shelf anchors have threaded interiors with knurled (roughened) exteriors. The internal threads allow fasteners to be easily removed, while the knurled exterior provides a rough texture that resists failure by pullout. Typically however, these anchors are designed to be press-fitted into the receiving holes. It is recommended instead that the anchors be adhered in place using a two-part epoxy. Ideally, anchors should be approximately 0.79 mm (1⁄32 in.) smaller in diameter than the hole to assure a sufficient adhesive bond line all around the circumference of the anchor (fig. 14).

If the opportunity exists, anchors can be machined from reverse-thread rod stock. The reverse threads on the exterior of the anchor allow the anchor to be removed by turning it clockwise and unscrewing it from the hole. Before insertion into the hole and application of the adhesive, the anchor should be thoroughly coated with a release agent. This will ensure that the adhesive will not fully bond to the surface of the anchor and that it will be easily removed if necessary.

Wall Mounts When the display of an object necessitates that it be mounted to a vertical surface, the mount must be designed in a manner that addresses unique needs. If the mount becomes the sole means of support, issues of strength, stiffness, and secure capture of the object become paramount. Typically, a wall mount consists of at least two interlocking parts: one fastened to the wall and one capturing the object (figs. 15a–d). Provision for an easily accessible mechanical union that eliminates any possible movement should be incorporated into the design of each element.

Accommodating Points for Anchoring When the object is anchored to the floor, wall, pedestal, or case, either by directly bolting it or by using a

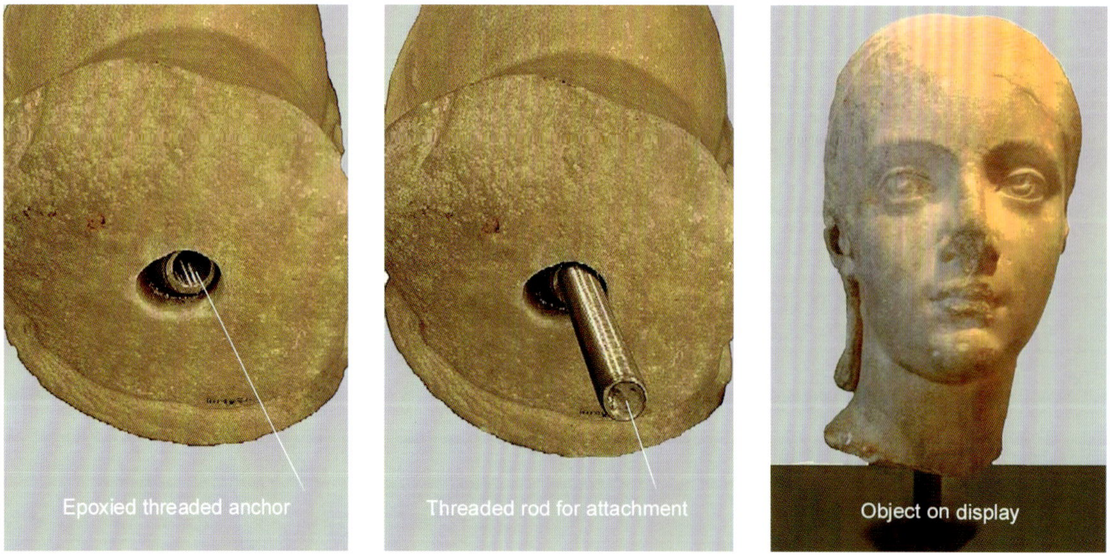

Figure 14
A threaded anchor adhered into a marble head. The support pin is then threaded into the anchor to support the object. *Portrait Head of Plautilla*. Roman, early third century A.D. J. Paul Getty Museum, 72.AA.118.

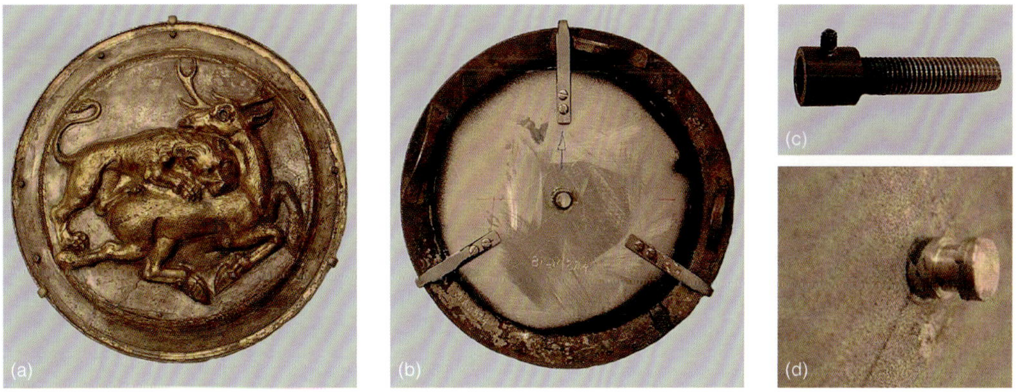

supporting/restraining mount, the object essentially becomes part of the building. Because of this condition, the anchor mechanism must be substantial and sound. A number of practical innovations to the building structure can help greatly to facilitate solid anchoring. Anchor points embedded in key areas of gallery floors can be used to bolt down pedestals and cases without the necessity of continually drilling anchor holes in the floor for each change of installation. At the Getty Museum at the Villa regular anchor points have been built into the floors, and when not in use, they are disguised by cover plates that are part of the overall floor design (fig. 16). The criteria used in the design of the anchor points established their capability of securely anchoring a 1451.5 kg (3200 lb.) object on a 122 cm (48 in.) high pedestal with a footprint of 76 cm by 76 cm (2.5 × 2.5 ft.), protecting it from the forces imposed by the design earthquake.

An anchoring (hanging) structure has also been added to the walls for securing cases, pedestals, or independently wall-hung objects of significant weight (fig. 17). A series of steel supports with regularly spaced groups of threaded holes (groups of four every 15 cm, or 6 in., along the length of each

Figures 15a–d
A silver phalera, shown as displayed (a) and from the back, with the mount structure exposed (b). The mount receiver anchors to the wall (c). The mount pin (d and b) is part of the central mount disk. The mount pin fits into the mount receiver and is locked in place with a set screw. *Phalera with a Relief of a Lion Attacking a Stag*. Parthian, second century B.C. J. Paul Getty Museum, 81.AM.87.4.

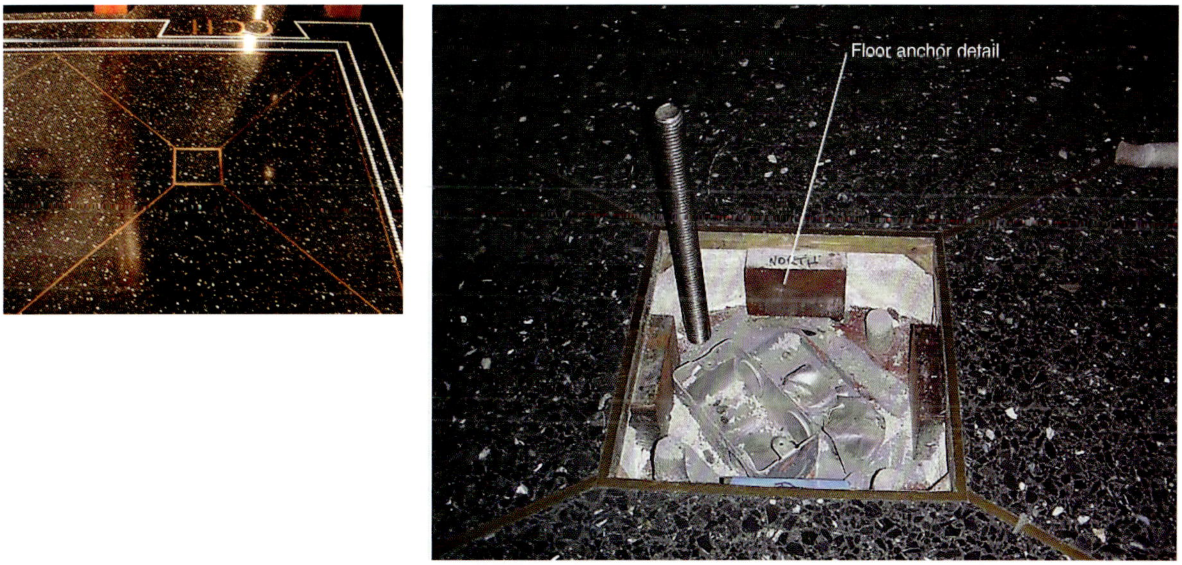

Figure 16
Gallery floor anchors.

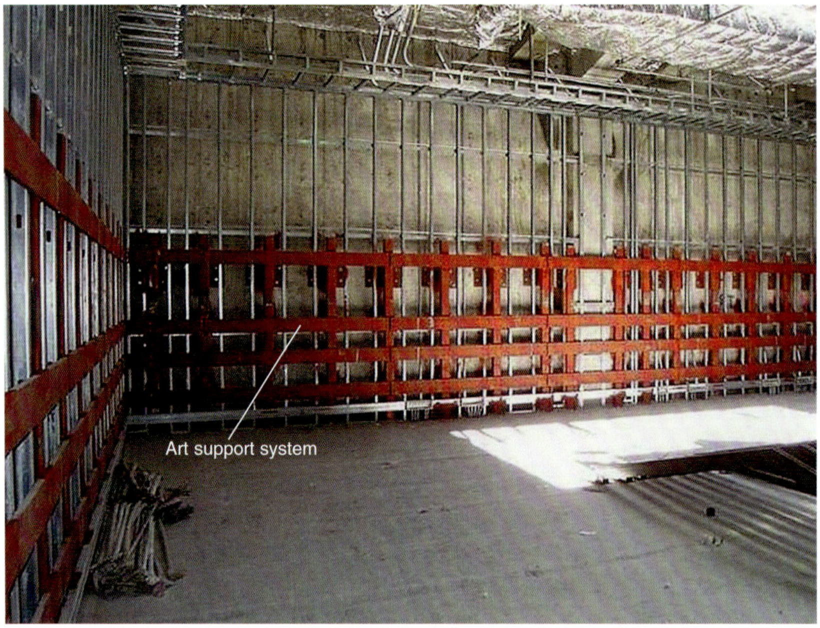

Figure 17
Steel flat stock 2.54 cm (1 in.) thick, with sets of regularly spaced threaded holes, shown during installation. The steel supports were installed at four heights to accommodate a wide range of hanging heights. The system was then covered over with finished gallery wall.

of the four levels) has been added to the gallery walls, behind the finished surfaces. The criteria used for designing the wall art support system established the capability of hanging a 362.8 kg (800 lb.) cantilevered object from two anchor points, a 1451.5 kg (3200 lb.) object from three anchor points (plus floor support), and a 544 kg (1200 lb.) flush-mounted relief on any two attachment points while it is subjected to the forces implied by the design earthquake. For example, the wall support system was effectively used to anchor large mosaics in an exhibition at the Getty Museum at the Villa (figs. 18a and 18b).

Decoupling: Seismic Base Isolation

Up to this point the discussion has been limited to the anchoring of objects with a variety of support mounts that essentially make the object part of the structure. While this approach has a number of advantages, it also means that the seismic forces transmitted through the building will be fully transferred to the object. It also requires either that the mount be fully visible in the display or that an internal structure (sometimes invasive to the fabric of the object) be introduced. Since this is not always possible and since objects can be too fragile to withstand even a portion of the seismic load, an alternative approach can be pursued—that of base isolation.

Techniques for the isolation of structures have developed rapidly, but the isolation of contents—whether they be medical, scientific, or computer equipment or works of art—has lagged behind the trend.

Relatively speaking, base isolation remains a new solution to the reduction of seismic forces. In general, the isolation mechanisms and materials on which a building or an object rests are designed to absorb the motions and energies of the earthquake. Isolation mechanisms that have some form of restoring force are widely recognized as the most effective. Decoupling, as an approach to seismic isolation, allows the floor under the object to move during an earthquake without transferring the full force of the earthquake to the object. In a sense, the

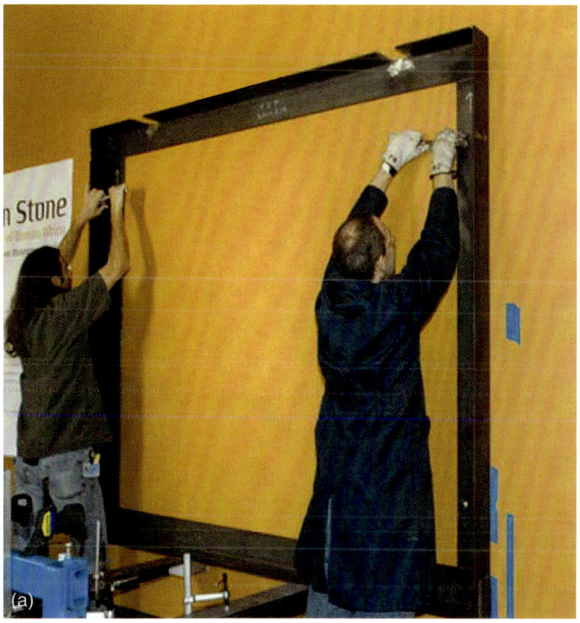

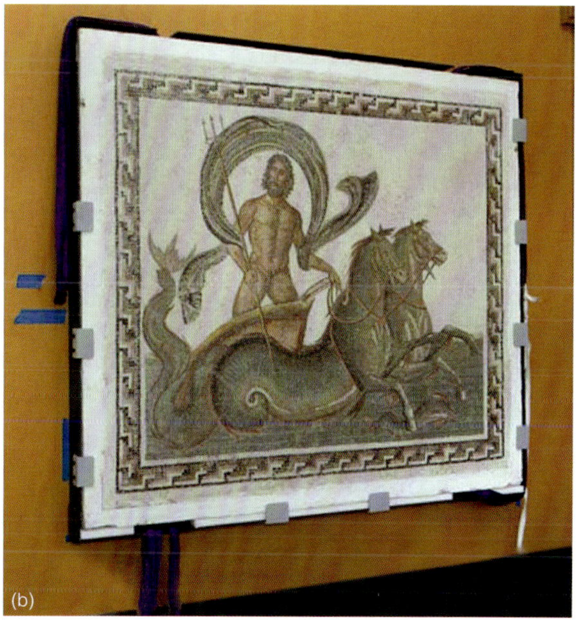

Figures 18a and 18b
Steel support frame for a mosaic panel, during installation (a) and installed (b). The installation onto the gallery wall utilizes two levels of the art support system shown in figure 17. For the installed panel, clips, attached to the steel frame, overlap and restrain the mosaic. *Neptune in a Chariot Drawn by Seahorses.* Roman, fourth century. Sousse Museum, Tunisia.

friction between the bottom of the object and the floor is eliminated or dramatically reduced through the introduction of low-friction interfaces or mechanisms that provide limited lateral movement between the object and the floor. There are a number of ways in which an object might be decoupled from the floor. As already described, early efforts to stabilize objects at the Getty Museum included altering the *B*/*H* ratios of pedestal-object assemblies by the addition of large steel plates to the bottom of the pedestals. These plates reduced the risk of overturning during an earthquake but did not, necessarily, stop rocking and the resultant dynamic pounding (rocking-induced impact) at the lower edges of the pedestal. Teflon pads were added to the underside of the plates to reduce friction. Theoretically these pads allowed the pedestal to slide further, reducing the overturning threat and minimizing the degree of rocking. In practice, however, this decoupling was imperfect, since rocking, even overturning, was made even more likely by encounters of the sliding pedestal with imperfections in the floor that dramatically, and suddenly, increased friction.

Using the site and building study done by Lindval, Richter and Associates (1984), which identified a maximum probable event and then provided design earthquake spectra, Getty Museum staff found that any isolation mechanism being considered would have to have a period of greater than 0.9 sec. to get any reduction of acceleration input, estimated to be 0.7 g at its greatest.

The longer the period of the isolator, the greater the isolation, as long as sufficient room for displacement is provided. However, at some point displacement demands would be impractical to accommodate, either for reasons of display aesthetics, limited square footage in the galleries, or safety of the visitors.

Based on the data developed from the design earthquake specific to the Getty Museum at the Villa, it was determined that differing degrees of protection can be achieved for the Getty collections by modifying certain aspects of the base isolator design. To achieve 60% isolation, the mechanism must be designed to have a 2 sec. period and to accommodate a minimum of 30.5 cm (12 in.) of displacement. This design results in the lower portions of the object being subjected to a peak horizontal acceleration of approximately 0.3 g. If the isolator is modified to accommodate 45.7 cm (18 in.) of displacement and designed for a period of 3 sec., the lower part of the object will be subjected to a peak acceleration of 0.2 g, which translates to a 70% isolation. In both cases, a 5% damping, introduced by the isolator mechanism itself, is assumed.

In 1985 Jack Yaghoubian of Quantec Systems in Los Angeles proposed an isolation mechanism based on an inverted dish system for the protection of a large marble sculpture in the collection. Shake table tests of the isolator's performance proved promising, and when the Museum's design earthquake spectra were applied, the isolator proved to significantly reduce the risk to the object.

Numerous other designs were evaluated by the Museum from 1985 to 1990. A design originating in the Museum's Department of Antiquities Conservation was tested at a commercial shake table in 1990 (sine dwell, random dwell, and simulated earthquake), and these tests indicated that the mechanism had a natural period of 3 sec., which, when combined with a 45.7 cm (18 in.) displacement capacity, provided an almost 70% reduction of the seismic forces at the top surface of the isolator. The shake table tests indicated that the isolator had a period of 2.4 Hz (approximately 4 sec.). This design provided an acceptable compromise between displacement demands and the size of the transmitted earthquake force.

The isolator is a three-level decoupling mechanism that offers relative displacement among the top, middle, and bottom platforms. The top and middle platforms are supported by orthogonal sets of captured linear bearings that travel along rails. The orthogonal arrangement of these rail-bearing supports prevents torsional movements. Forces arriving at the isolator from a diagonal orientation are accommodated by a lateral "scissoring" action of the upper and middle platforms with respect to one another. The bottom frame is attached rigidly to the floor, while the upper frame provides an attachment level for the pedestal, case, or object. Vertical restraint is achieved through the mechanical capture of the linear guide blocks to the rails they travel on. As a result, there is no opportunity for uplift of the assembly during an earthquake. It should be noted that since 1984, some assumptions have changed in seismic engineering, in particular because of data provided by subsequent earthquakes—especially the 6.7 Richter scale magnitude Northridge Earthquake of January 17, 1994, which resulted in a 0.25–0.50 g lateral acceleration and a vertical record of 0.19 g in the Los Angeles area. The more recent events have led to a reevaluation, carried out in 2005, of the design parameters for the Getty Villa site. As a result of this new study, vertical capture is given greater attention than in the past (URS Corporation 2005).

Lateral displacement of the top and middle platforms is individually limited by a centrally mounted roller for each platform that travels along an angled ramp, compressing a series of springs that provide both a predetermined resistance to the lateral displacement and a centering force to the platforms. Spring rates are predetermined to provide a natural period between 1 and 3 sec.—the variation being determined by the available displacement. A springs-in-series design provides two specific ranges of resistance to the lateral motion.

A softer set of springs provides a longer period with less resistance. As maximum displacement is approached, the stiffer set of springs offers greater resistance in order to accommodate larger earthquake forces and to prevent a sudden stop as the maximum displacement is reached.

During the 1990 testing the isolator was attached to a full-scale model of the object being considered for exhibition. The weight distribution of the model accurately mimicked that of the original sculpture, as did the approach to the assembly of the object's fragments and the attachment to the base and isolator (figs. 19a–c). The 100% design earthquake (maximum probable event) motion was filtered to remove periods greater than 4 sec., to ensure that the maximum displacement of the table would not be exceeded. Although some whipping at the top of the sculpture model was observed, the top of the sculpture displaced with a maximum excursion of 6.3 cm (2.5 in.). The predicted 0.7 g peak was reduced to 0.1 g at the top of the isolator (a 70% reduction), to 0.15 g at the top of the pedestal (a 60% reduction), and to 0.3–0.4 g (a 35%–45% reduction) at the top of the sculpture.

Since the isolator was designed for a full 45.7 cm (18 in.) displacement, ample reserve was provided by the design. Realistically, however, this

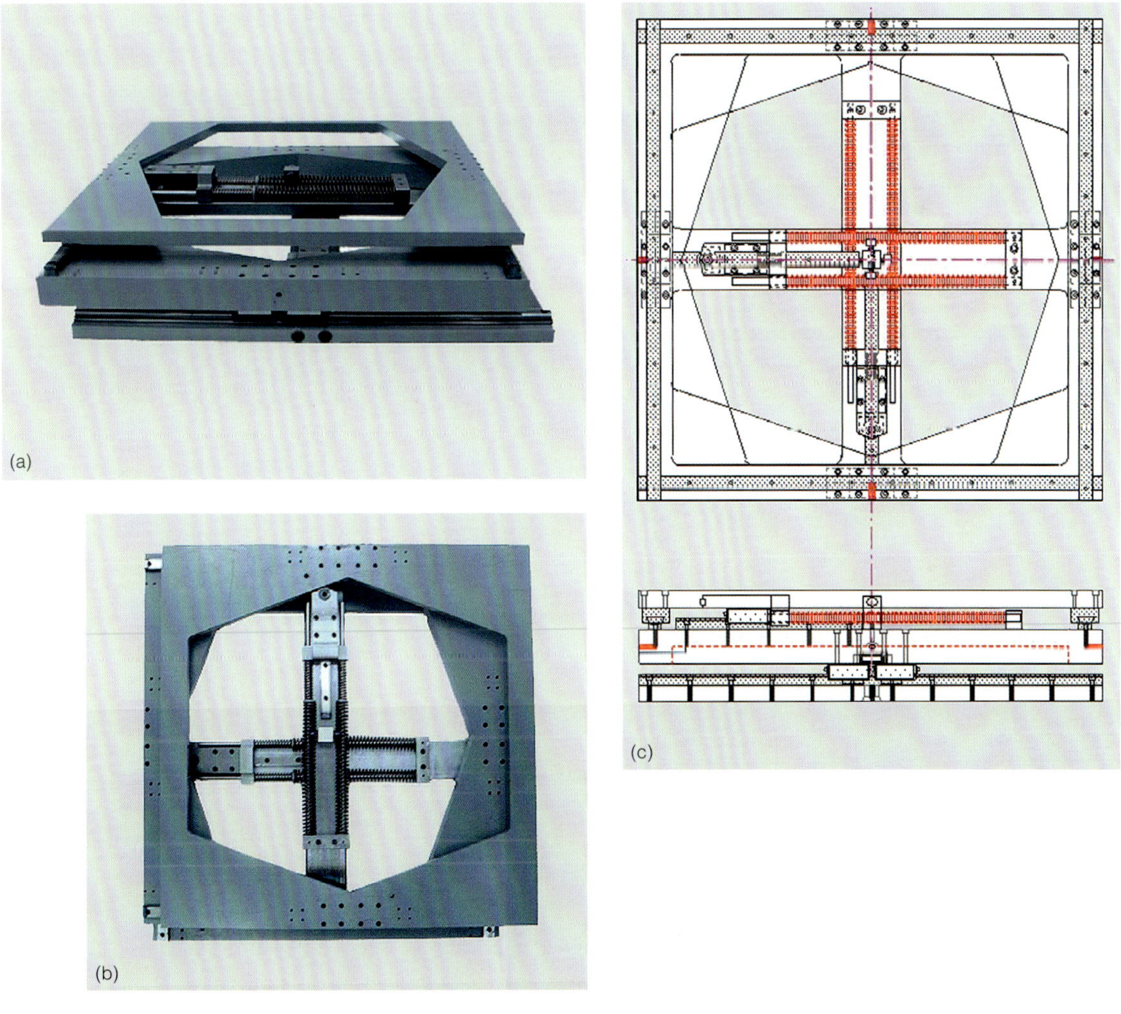

Figures 19a–c
The isolator unit used at the Getty Museum at the Villa. Photographs from the side (a) and from the top (b) show some displacement in both axes. Diagrams (c) show the plan view (top) and the elevation (bottom).

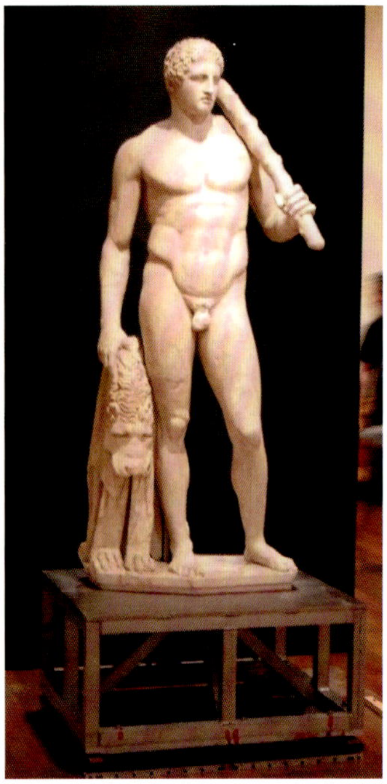

Figure 20
An aluminum cage construction supporting a large marble sculpture, assuring a rigid and strong support. The frame is then covered with a facing for display purposes. *Statue of Hercules (Lansdowne Herakles)*. Roman, ca. 125 A.D. J. Paul Getty Museum, 70.AA.109.

amount of displacement is not always possible because of the limitations of gallery space, aesthetic proportions of the pedestal to the object size, and visitor safety, should the isolator and ground displace while the visitor is standing nearby.

It should be noted that while isolators absorb a given percentage of the seismic forces, they can never eliminate the need for seismic mounts and structurally robust exhibition furniture.

Casework

The structural design of exhibition casework is a critical component in any effort to mitigate seismic damage to exhibitions. Display cases and pedestals must be sufficiently stiff and structurally strong, designed to withstand dynamic forces beyond the predicted event while remaining intact and securely anchored to the building. As a general rule, designing for a force of 3 g (which includes a safety factor of 3 or more) meets a wide variety of needs.

The design of a case or pedestal structure should provide direct support under the display surface and artwork (fig. 20). This support structure should be rigidly connected to the structural elements of the case or pedestal. The casework structure should include attachment points either to the building or to an isolation system.

Anchoring Casework Anchoring the casework to the floor or wall is always the best choice, since this procedure fixes the display furniture firmly in place and results in a synchronous movement of the casework with the building. It is important, however, that these anchoring points and the hardware used be sufficiently strong to resist the forces imposed on them during an earthquake (fig. 21).

Display Accessories Casework accessories, such as risers, are part of the link between the artwork and the display furniture. Their design must accommodate a dynamic load, and depending on the weight of the artwork, extra reinforcement may be required (figs. 22 and 23). It is very important that the riser have a firm connection to the display surface. For lighter objects, this

Figure 21
Points of attachment of the pedestal to the floor and of the object to the pedestal. They are both easily accessed in this pedestal through a removable side panel.

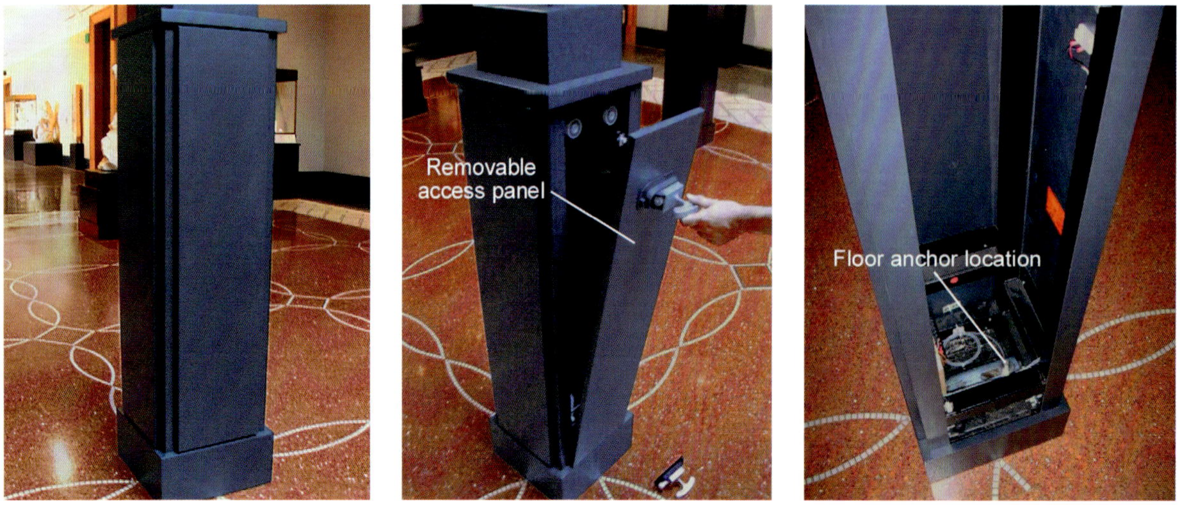

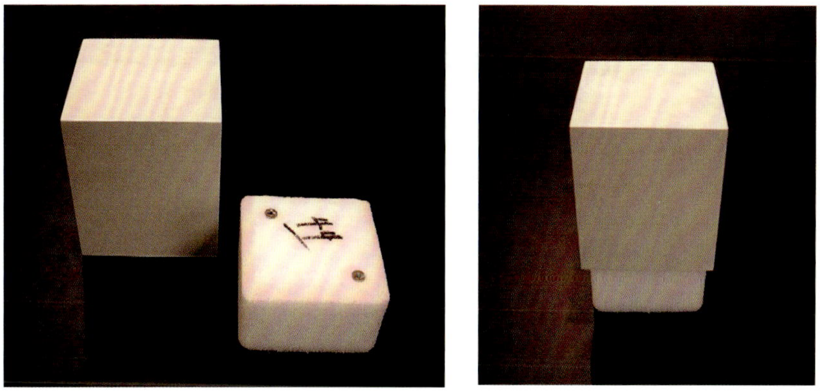

Figure 22
A foam insert support block anchored to the deck. A hollow buildup that will support the object (normally small and lightweight for this type of support) is slipped over the foam block.

goal can be easily achieved by affixing a block of extra-firm Ethafoam, which fits intimately inside the riser, to the display surface. Once the artwork is installed on the riser, it can be fitted over the foam block. A tight fit between the foam block and the riser is crucial.

It is imperative to have a solid connection among all the casework components, including display backboards, light hoods, and louvered panels. Anything that could come loose within the casework should be secured.

Conclusion

Many of the suggested solutions outlined in this paper for the protection of collections from seismic threats have concentrated on exhibition conditions. It should be kept in mind, however, that the majority of objects in many collections are not always on display in the galleries but, rather, placed in storage areas where the threat of seismic damage can be just as great, if not greater (because of density). The concepts of mitigation presented here work equally well for storage facilities. Indeed, when they are applied with less concern for aesthetic presentation but with equal concern for access and safety, they can be

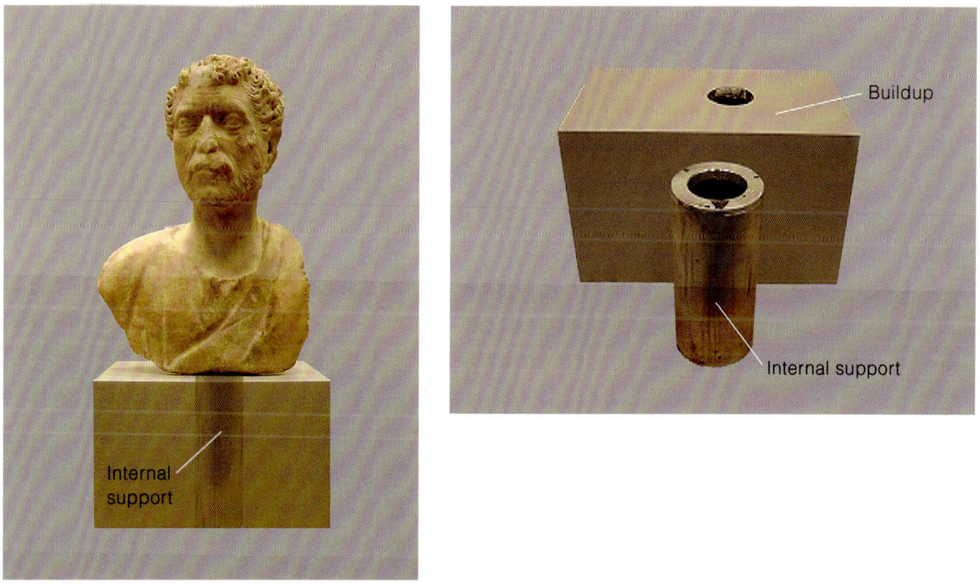

Figure 23
A riser with an internal support. For heavier objects, fastening the riser directly to the display surface is advisable. In some situations an internal support within the riser might be needed to transfer the weight of the artwork to the structural deck of the exhibit case or pedestal.

far more straightforward and considerably less expensive. Tying objects to shelves that have been firmly secured to the wall, placing large restraining lips or ledges along the length and outer edges of shelves, placing soft buffering foam pads between objects in proximity, or carving individual cavities in large blocks of Ethafoam for storage are all effective ways of protecting stored collections. None are excessively expensive, and they do not necessarily require extensive engineering studies to carry out.

The Getty Museum has pursued the protection of its collections from earthquake damage for several decades. With the assistance of informed engineers and seismologists, the mount makers, conservators, and preparators have arrived at a number of effective ways of achieving their preservation goals. The effort continually evolves, as does seismology and our understanding of the nature of earthquakes. It is the authors' hope that the efforts and solutions described in this paper will be advanced by the interested and involved reader, and that as a result, fewer collections will suffer from the destructive forces of the inevitable earthquakes yet to come.

References

Agbabian, M. S., S. F. Masri, and R. L. Nigbor
1990 *Evaluation of Seismic Mitigation Measures for Art Objects* (Los Angeles).

Lindvall, Richter and Associates
1984 "Geologic and Seismic Safety Evaluation of the Main Museum Building for the J. Paul Getty Museum, Malibu, California," Department of Antiquities Conservation, J. Paul Getty Museum, Los Angeles.

Podany, J. C., and P. T. Leavengood
1992 "Chinese Monumental Sculpture: Treatment and Seismic Stabilization," *Proceedings of the Seventh International Congress on Deterioration and Conservation of Stone* (Lisbon).

URS Corporation
2005 "Ground Motion Response Spectra and Time Histories, Getty Museum, Malibu," Department of Antiquities Conservation, J. Paul Getty Museum, Los Angeles.

Finalizing the Design of Seismic Isolators for the Antiquities Collection of the J. Paul Getty Museum

Matthew J. Schoettler

Andreas Stavridis

Introduction

The J. Paul Getty Museum at the Villa is located in Malibu, California. The recently renovated Getty Villa houses the Museum's antiquities collection, which includes substantial works of art with origins in ancient Greece and the Roman Empire. The Museum is located in an area of high seismicity, with major faults within a distance that poses a threat to the antiquities. Earthquakes of the last two decades, such as the 1994 Northridge Earthquake, with an epicenter 19 km (12 mi.) from the Getty Villa, emphasized the seismic threat. Aware of the risk, the Getty Museum has made the seismic protection of the collection a priority.

To reduce the risk of seismic damage to its antiquities, the Museum initiated a case study involving object-specific base isolators. This study updated the mechanism and seismic design of seven existing isolators used to protect some of the collection's most vulnerable pieces. The antiquities had been provided with base isolation as part of a seismic mitigation effort started by the Museum over twenty years ago. However, advances in seismology and earthquake engineering since that time led the Museum to initiate this recent study to update the design and revise the dynamic properties of its isolators according to current earthquake engineering knowledge.

The isolated antiquities, listed in table 1, encompass a wide range of materials, sizes, and weights. The antiquities are made of various materials, ranging from marble statues to a terracotta amphora, and their weights vary between 2.2 and 408.6 kg (5–900 lb.) These artworks were selected to be provided with base isolators because of their historical significance coupled with their seismic vulnerability, due to their geometry and structural defects caused over the centuries. A design specific to each individual isolator was required to address the differences in displacement capacity and isolated mass.

The objective of this study was to lower the seismic forces each of the isolated antiquities would undergo in a seismic event. This goal was met through a seismic design procedure that involved initial analysis, preliminary design, verification testing, and a final design. The final design was based on the seismic demand evaluated for two risk levels that were estimated through a probabilistic seismic hazard analysis (PSHA). However, no performance criteria related to forces were accounted for in the design, as the capacities of the antiquities to withstand earthquake forces are not known. In light of the unknown

Table 1
Object and isolator properties (measurements are given in metric units).

Object	Object weight (kg)	Isolated weight (kg)	Exterior dimensions (cm)	Displacement capacity (cm)
Kouros	408.0	713.0	119.0 × 119.0	±34.0
Orpheus Group	195.0	637.0	140.0 × 196.0	±35.5
Pouring Satyr	272.0	387.6	61.0 × 61.0	±23.5
Attic Panathenaic Amphora	7.7	116.0	62.5 × 62.5	±23.5
Cypriot Fertility Goddess	11.0	68.5	48.6 × 48.6	±17.8
Male Harp Player	13.6	74.8	69.2 × 69.2	±17.8
Glass case	2.2	109.7	50.0 × 50.0	±11.4

capacities, the objective of lowering the seismic forces to the extent possible, given the displacement capacity of the existing isolators, was established.

Base Isolation Background

Base isolation reduces the inertia forces acting on a structure during seismic events. This seismic mitigation technique is used in important civil structures, such as hospitals and highway bridges, and is increasingly being used in the field of museum preservation. Whether this technique is implemented for a building, bridge, or museum artifact, the fundamental concept is the same.

The isolator provides an element of reduced stiffness that lowers the frequency of the system. A shift in frequency can be regarded as an inverse shift in terms of period, as shown in figure 1. Thus, the reduced stiffness introduced by the isolator corresponds to a larger period that is outside the range of periods that contain large acceleration demands.

A stiff object fixed to the floor would be subject to large accelerations, but relatively small deformations would be induced. This is observed in the response spectra presented in figure 1, where a stiff structure or object has a low period. The solid line identifies the acceleration demand with respect to the system's period, while the displacement demand is indicated by a dashed line. A stiff object is required to resist large inertia forces, while the displacement demand is comparatively low. As the period increases, the acceleration demand reduces, and the displacement demand increases. The response, in terms of acceleration, of a flexible structure with a period in the range of 2–4 sec. is notably lower than that of a stiffer structure. However, the deformations imposed by the earthquake are significantly larger. This trade-off between reduced inertia forces and increased displacement is characteristic of base isolation.

The selection of an isolated period depends on the appropriate response spectra that represent the intensity of a potential earthquake. The distance to the tectonic faults, the location and local site conditions, the accepted risk level, and the system's damping level have pronounced impacts on the response spectra curves. Therefore, the response spectra need to be generated to reflect a unique risk level at a specific site.

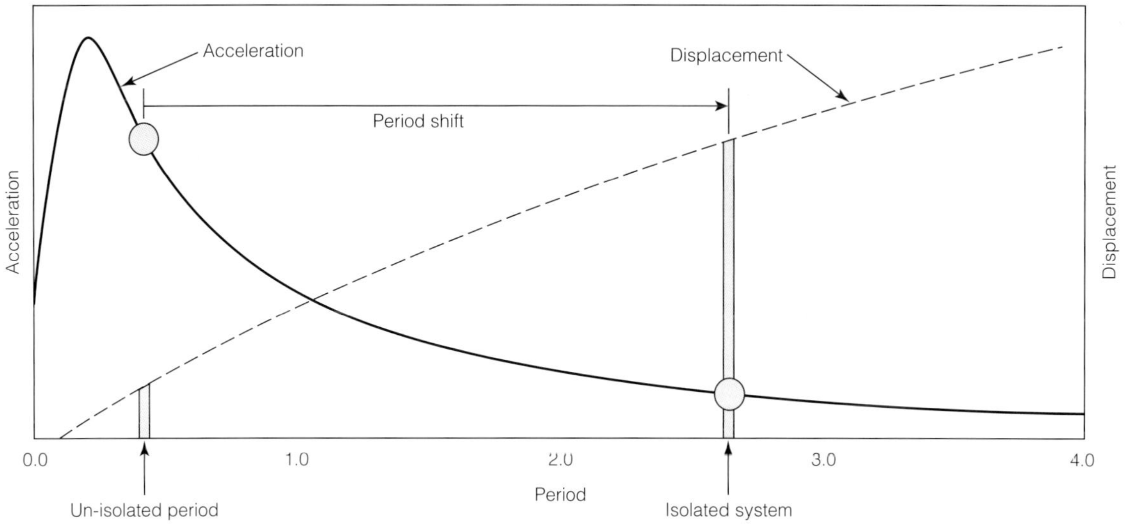

Figure 1
Comparison of acceleration and displacement response spectra. Photo: Matt Schoettler.

Isolator Description

The mechanism implemented in the Getty Villa's isolators introduces modifications and improvements to earlier mechanisms and is the result of efforts carried out by the Getty Museum over the last twenty years. Further modifications were suggested and incorporated as a result of the tests carried out during this study. The isolators' dimensions were selected by the Museum's staff in accordance with the constraints imposed by the museum setting prior to the most recent seismic demand evaluation. The isolators were assembled in-house by Getty Museum staff and varied in terms of their dimensions and displacement capacities.

Table 1 identifies the in-plane dimensions of the isolators and their corresponding displacement capacities. It is evident that the displacement capacity is dependent on the overall dimensions of the isolator, as a wider isolator allows for a larger displacement capacity.

The isolators consist of three stacked metal frames capable of displacing in two uncoupled perpendicular directions. The restoring force is provided by compression springs acting between two adjacent frames, as illustrated in figure 2 and figure 3a. Figure 3b depicts the principle of operation of the isolator mechanism. Springs are installed in parallel on two guide rods at each of the two levels. The springs act against a reaction block and the roller block. The isolating mechanism is a passive system not requiring a computer or electricity to operate. Inherent damping is the only form of energy dissipation, as no external damping mechanism is provided. The overturning moment is resisted by the sliding bearings connecting adjacent frames. Further description of the isolator devices can be found in the paper by Lowry, Farrar, Armendariz, and Podany in the present volume.

This study assumed that each of the isolators would be rigidly attached to the floor and that the objects would be firmly attached to the isolators. A connection to the floor capable of withstanding the forces developed by the isolator is required to prevent sliding and overturning so that the isolator can perform as intended. The implemented base isolation mechanism still requires

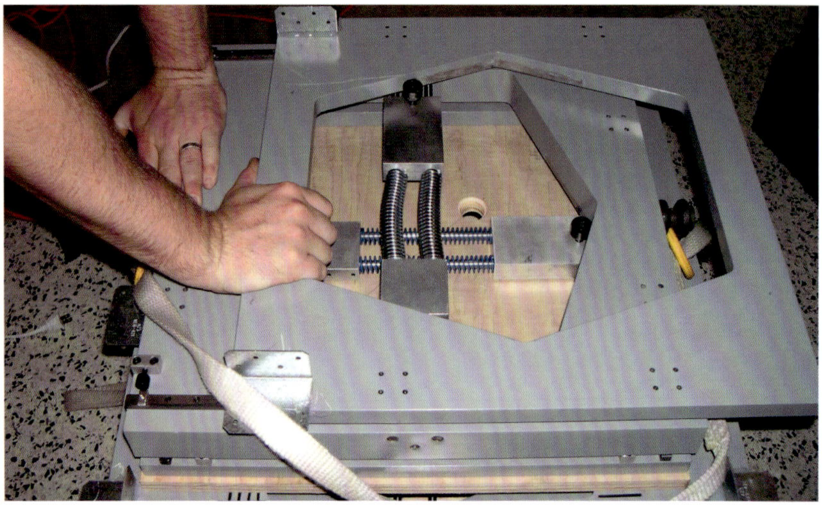

Figure 2
Displaced isolator configuration. Photo: Andreas Stavridis.

that the antiquity be mounted to the deck, because seismic forces are reduced but not eliminated. It is noted that the vertical component of excitation is not isolated, because of the sliding bearings used to restrain uplift. The forces transferred to the objects in the vertical direction are accounted for when the Museum's mount makers design the mounts for the isolator.

These base isolators possess desirable characteristics from the perspective of a museum. Because of their small heights, they can be easily concealed in cabinetry, which is an attractive feature for installation in museums. The most intriguing feature is that the devices are not object specific. An isolator can be easily tuned for a different object by replacing the springs with others suitable to the characteristics of the new object. This flexibility could prove useful if the priorities for the isolated artifacts change. For instance, the artifacts displayed in a gallery devoted to visiting exhibitions could use the existing isolators if the

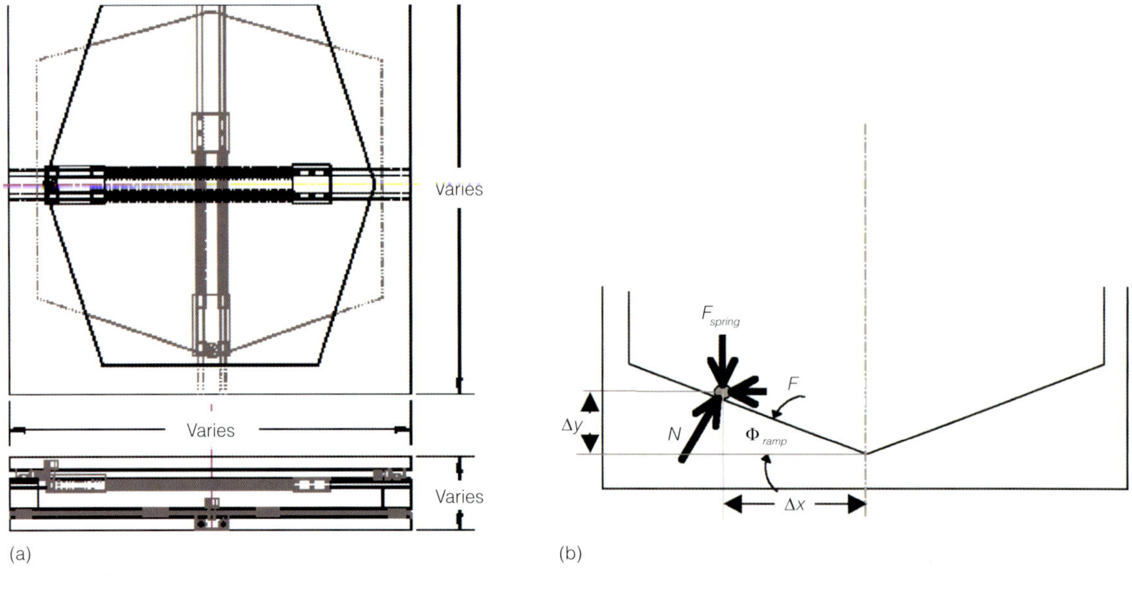

Figures 3a and 3b
The isolator mechanism, shown in a plan drawing of the isolator frame (a) and a diagram of the principle of operation (b).

necessary adjustments were made. The needed adjustments involve a calculation of the required properties of a new set of springs, such that the stiffness of the system is changed to account for the change in mass. The procedure would be more complicated if the strengths of the different objects were also to be accounted for.

The base isolation mechanism permits a direct and relatively simple design procedure from an engineering standpoint. Analysis can be conducted assuming a single-degree-of-freedom structure. A linear elastic response is implemented if linear elastic compression springs are used to provide the self-centering restoring force. This procedure allows for an efficient method of estimating the response to a specific hazard level from the appropriate response spectrum.

In the Getty Villa isolators, however, a modification was studied. This modification involved the placement of two springs in series around each of the guide rods. The main difference introduced with the springs-in-series setup was a force-displacement relationship that became bilinear although it remained elastic. The bilinear response is obtained with the use of a stiff spring placed in series with a soft spring, and it is qualitatively illustrated in figure 4. To obtain a force-displacement curve of that type, the soft spring is required to completely compress at a displacement level lower than the isolator's displacement capacity. Ensuring an accurate soft spring's compressed length is essential for this spring setup to perform as intended. The bilinear portion of this response introduces a significant increase in the system's stiffness that is activated when the isolator displacement reaches about 90% of its displacement capacity. This desired increase in stiffness acts as a safety mechanism that lessens the likelihood of exhausting the displacement capacity. This point is very important, since if the displacement capacity is exhausted, the sliders would impact the stops and the object would be subjected to a high level of forces. This change in the design reduces the velocity of the antiquity as it approaches the displacement limit and reduces the likelihood and the consequences of a possible impact. Although this design is more complex, it was implemented in this study because of the low displacement capacity of some of the isolators compared to the estimated seismic demand at the specific site.

The parameters defining the bilinear response are the soft spring stiffness, K_{soft}, the stiff spring stiffness, K_{stiff}, and the displacement at which the soft spring fully compresses, δ_c. The soft spring dominates the response for most of

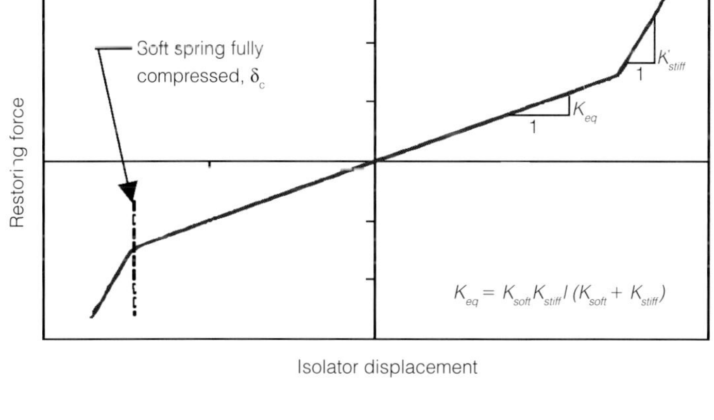

Figure 4
Force-displacement relationship.

the displacement trajectory. In figure 4, this region is the portion of the force-displacement curve within $\pm\delta_c$ with a stiffness K_{eq}. This stiffness is used for the calculation of the initial period of the isolator, since the response is controlled by the equivalent stiffness of the springs in series. The initial period is indicated as the design period. This response does not exhaust the compression capacity of the soft spring and is activated for moderate earthquakes. During the design-level earthquake that has a return period of 475 years or, equivalently, a probability of exceedance of 10% in 50 years, the soft spring will completely compress when the displacement demand approaches the displacement capacity. Compression of the soft spring results in a drastic increase in stiffness that hinders the exhaustion of displacement capacity. In figure 4, the increased stiffness is identified as $K_{stiff,}$ and it equals the stiffness of the stiff spring, which is the only spring acting for the remainder of the displacement trajectory. The likelihood and effects of a possible impact are reduced through this bilinear design.

Analysis and Initial Design

The initial analysis included the derivation of an equation for the restoring force as a function of the isolator displacement. This force-displacement curve is needed to obtain the isolator's stiffness, which is essential to calculate the period. In figure 3b, the forces acting on the roller at a displaced position are illustrated. This relationship is obtained as follows:

$$F_{spring} = K_{spring}\,\Delta y \tag{1}$$

$$\Delta y = \tan(\varphi_{ramp})\,\Delta x \tag{2}$$

$$F = \tan(\varphi_{ramp})\,F_{spring} \tag{3}$$

where F_{spring} is the equivalent spring force, K_{spring} is the acting spring stiffness, Δy is the spring compression, φ_{ramp} is the ramp angle, Δx is the lateral displacement, and F is the lateral restoring force, as demonstrated in figure 3b.

If equations 1 through 3 are combined, the isolator force-displacement relationship can be obtained as follows:

$$F = (K_{spring}\tan^2\varphi_{ramp})\,\Delta x \tag{4}$$

The resulting isolator stiffness is

$$K_{iso} = \frac{F}{\Delta x} = K_{spring}\tan^2\varphi \tag{5}$$

Thus, the isolator period (T_{iso}) is given as

$$T_{iso} = 2\pi\sqrt{\frac{W}{g\cdot K_{iso}}} = 2\pi\sqrt{\frac{W}{g(K_{spring}\tan^2\varphi_{ramp})}} \tag{6}$$

where g is the acceleration of gravity and W is the weight of the isolated mass. The isolated mass includes the frame(s), the antiquity, and any other cabinetry or vitrine above the isolated frame. When springs in series are implemented, with a soft spring and a stiff spring installed on each guide rod, equation 6 is used with the acting spring stiffness being equal to the equivalent spring stiffness.

Free Vibration Testing

To obtain a reliable estimation of the dynamic properties, pullback tests were conducted. The objectives of these tests were the validation of equation 6 as well as the measurement of the damping ratio of different isolators at a range of periods. Preparations for the tests included the acquisition of masses to simulate the weight of the antiquities and the calculation of properties for several spring configurations. The sets of springs were obtained through the application of equation 6 and resulted in isolator periods from 1 to 4 sec. This was identified as the period range of interest based on the design response spectra. For the tests, a substitute mass representing the antiquity, decking, and glass vitrine was attached to the isolator. The isolators were displaced and instantaneously released with the use of a quick-release mechanism, so that free vibration motion was initiated. Two unidirectional accelerometers, firmly attached on parallel sides of the moving isolator frame, were used to record the motion during each test.

Three isolators of different sizes were tested in order to identify the damping trends, and the results were then extrapolated to the remaining isolators. Within the period range of interest, the relationship between the damping ratio and system's period was identified, so that the appropriate damping level would be used in the final design. The investigation included the comparison between two arrangements—one with a single spring per guide rod and the other with springs placed in series around each guide rod. The latter configuration provided the intended elastic, bilinear force-displacement relationship.

With the structure behaving elastically, the damping calculation was based on the logarithmic decrement formula (Filiatrault 2002), as the viscous damping formulation matched the tests results better than the coulomb-type friction damping (Chopra 2001). Figure 5 presents the acceleration time histories recorded during one of the free vibration tests. This test yielded a 1.5 sec. period and a 7.5% damping ratio. Figure 5 also shows the analytically calculated, exponentially decaying, sinusoidal line with the measured period and damping ratio, thus validating the extracted dynamic properties. The agreement between these measurements demonstrated that a viscous damping

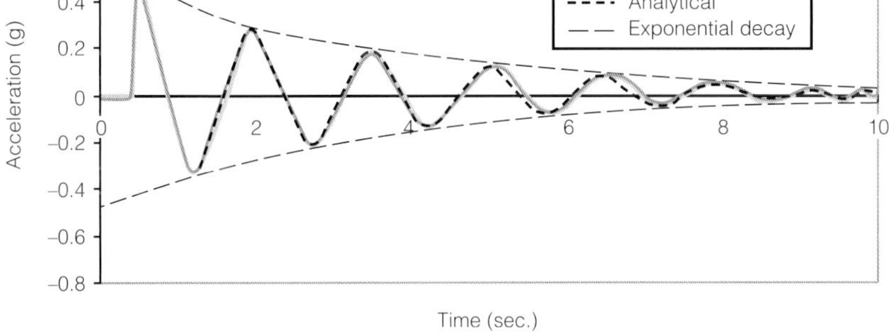

Figure 5
Free vibration of the middle frame of the Attic Panathenaic Amphora isolator with a period of 1.5 sec. and a damping ratio of 7.5%.

model would be acceptable for use in the time-history analyses employed to optimize the final design.

The damping values obtained from all tests ranged from 5% to 14%. Damping test results are plotted with respect to the structural period in figure 6 for the isolator that was finally used for the Attic Panathenaic Amphora. It is interesting to note that damping values tend to increase for higher periods, for the top frame and for the cases of springs in series. This trend can be attributed to the rolling mechanism, to differences in isolated mass, and to the yielding of springs when they are fully compressed, but further study is necessary to evaluate the exact influence of each parameter. A second objective of these tests was the validation of equation 6, and this goal was satisfactorily met. The agreement of analytical prediction and experimentally measured values for the period is reasonable, as the maximum deviation from the analytical prediction is within 8% in all cases. Therefore, it is safe to assume that the derived force-displacement relationship adequately captures the actual behavior and can be used for the final design.

Final Design

An essential step in the design process is the determination of the seismic demand. An up-to-date estimation of the seismic demand is necessary for the design of an isolation system. Therefore, a PSHA was conducted for the site (URS Corporation 2005). The PSHA included developments in seismology that have emerged since a previous study (Lindvall, Richter and Associates 1984), such as rupture directivity, updated fault models, and ground motion attenuation relationships (Abrahamson and Silva 1997; Somerville et al. 1997; Somerville 2003). The PSHA carried out for the Getty Villa site estimated the seismic demand at two risk levels. The demand was presented in the form of acceleration response spectra at both 50% probability of exceedance in 50 years (72 year return period) and 10% probability of exceedance in 50 years (475 year return period). It was decided by the Museum that the design level should correspond to the more demanding hazard with a return period of

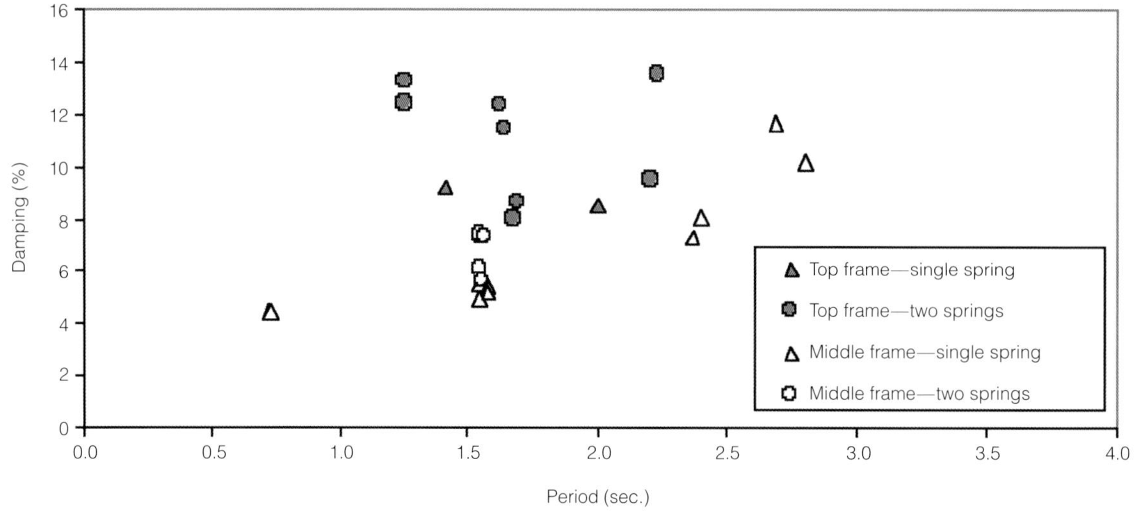

Figure 6
Free vibration test results for the isolator of the Attic Panathenaic Amphora.

Earthquake	M_w	Station	Distance (km)	Site	Scale factor
Loma Prieta	7.0	Corralitos	3.4	C	1.449
		San Jose Santa Teresa Hills	14.0	C	1.439
		Gilroy #6	18.3	C	0.851
Northridge	6.7	Rinaldi	7.1	C	1.482
Chi-Chi	7.6	TCU074	13.5	C	3.494
		TCU075	1.5	C	4.254
		TCU129	1.2	C	0.518

Table 2
Ensemble of ground motions with scaling factors that represent a seismic event with a return period of 475 years (measurements are given in metric units; M_w = earthquake magnitude; site C refers to a site classification based on soil properties).

475 years. Seven ground motions appropriate for the Villa site were provided for each hazard level, based on findings of the PSHA (URS Corporation 2005). The recorded ground motions were selected to satisfy the magnitude and distance relationships resulting from the disaggregation of the hazard spectra and were accordingly scaled. Table 2 identifies the seven ground motions used for the event with a 475 year return period.

The design of the isolators was driven by their respective displacement capacities and the estimated displacement demand. The displacement-based design aimed to bound the structural response within the isolator's displacement capacity while reducing the forces transmitted to the antiquity. As a starting point in the final spring selection, the isolator period, T_{iso}, was selected by matching the design displacement response spectrum with the isolator's displacement capacity. This period was then adjusted through an iterative procedure to account for the bilinear response.

The bilinear response of springs in series necessitates nonlinear time-history analyses to optimize the spring configuration. For the final design, a structural analysis framework was used to assess the isolator's dynamic response under earthquake loading for several bilinear configurations. The ground motions were scaled to match the design displacement response spectrum at the value of T_{iso}, as shown in figures 7a and 7b. Based on the displacement demands from the time-history results, the period was adjusted, the ground motions were rescaled to this period, and the analyses were repeated. The procedure was iterated until an acceptable response was achieved. After the final design was obtained, the response to seven ground motions with a return period of 72 years was also evaluated. This assessment was made to ensure that the isolators responded in the elastic range for these moderate seismic events.

Dynamic time-history analyses were conducted with the isolator modeled as a single-degree-of-freedom structure. The damping values used in the analyses were based on the free vibration tests; they can be found in table 3. The nonlinear behavior of the springs-in-series configuration was captured in the hysteretic model. The total isolated mass, referenced in table 1, was assigned to the single degree of freedom. The mass distribution and response of the antiquity were not taken into account, as this level of detailing was considered

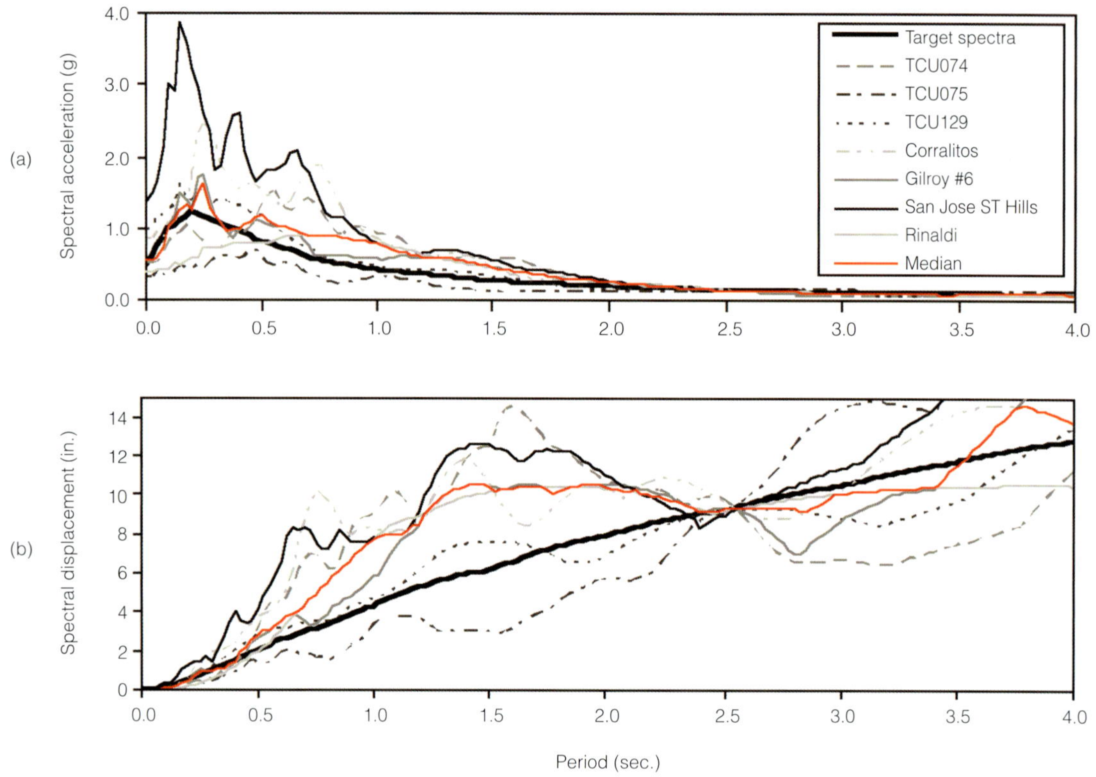

Figures 7a and 7b
Response spectra for a damping ration of 8% scaled to a 2.6 sec. period, in terms of acceleration (a) and displacement (b).

by the Getty Museum to be outside the scope of this study. The top and middle frames were modeled and assessed separately. According to a previous study, the Museum building has a fundamental period of 0.1 sec. (Lindvall, Richter and Associates 1984). The high stiffness of the Museum's structure allowed the analysis to be undertaken without consideration of the structural response of the building. The isolators were therefore assumed to be rigidly connected to the ground level where the input motion was applied.

Certain final configurations resulted in response records that exceeded the displacement capacity of the isolators. It was allowed that the demand exceed the capacity due to the variability provided by the seven ground motions. However, no more than two records were permitted to exceed the displacement capacity for a given isolator. Additionally, the median displacement demand of the seven time histories for both frames was kept below the displacement capacity for each isolator. In some instances, high scaling factors in conjunction with the shape of the response spectra penalized the springs-in-series configuration when it entered the stiffened region of the response. Although the high scaling factors resulted in unrealistic ground motions, the overall results supported the use of springs in series, as the median response was in all cases satisfying. Therefore, the bilinear elastic behavior was adopted, and two springs per guide rod were installed in the isolators.

The outcomes of the design procedure for the isolators were periods that ranged from 1.49 to 3.97 sec. Table 3 summarizes the design damping ratios and periods for each isolator resulting from the differing displacement capacities. Figures 7a and 7b present the scaled response spectra of the

Object	Displacement capacity (cm)	Isolated period (sec.)		Damping ratio (%)	
		Top frame	Middle frame	Top frame	Middle frame
Kouros	±34.3	3.97	3.83	12	12
Orpheus Group	±35.5	3.73	3.74	12	12
Pouring Satyr	±23.5	2.72	2.73	10	10
Attic Panathenaic Amphora	±23.5	2.55	2.56	8	8
Cypriot Fertility Goddess	±17.8	2.00	1.90	11	9
Male Harp Player	±17.8	2.01	1.91	11	9
Glass case	±11.4	1.49	1.49	8	7

Table 3
Final design parameters (measurements are given in metric units).

selected ground motions scaled at a 2.6 sec. period, which was the design period of the Attic Panathenaic Amphora (Stavridis et al. 2006). The scaled records, because of peaks and troughs, deviate from the design displacement spectrum at other period ranges, resulting in highly demanding input motions. The peaks at shorter periods become an issue when the soft spring fully compresses. For the remaining displacement trajectory, the isolator stiffness significantly increases and the response spectra can no longer predict the demand because of the nonlinear behavior. Figures 7a and 7b show that the response spectra of some scaled records indicate significantly high demand because of the scaling method.

The adopted design accounts for two levels of ground motion similar to the design of more traditional structures. It ensures that moderate earthquakes, with higher probability of occurrence, will excite only the initial range of the force-displacement relationship represented in figure 4, and the antiquities will be subjected to relatively low forces. Additionally, in case a more extreme event occurs, the isolators are less likely to run out of displacement and impact the adjacent frame. However, even if the earthquake demand could not be accommodated, the velocity of the isolator would be reduced, and the effect of an impact would be diminished compared to a single-spring design.

Suggested Design Procedure

A suggested design procedure is outlined in the flowchart (fig. 8). Important aspects of the design are presented as a template for future isolation efforts for important objects. In a museum setting, geometric and aesthetic preferences can have important impacts on the isolator dimensions. These constraints, as well as the layout of the objects in the gallery, must be considered prior to dimensioning the isolator, so as to achieve optimal seismic performance.

Summary

This paper has presented an overview of the study on the dynamic properties and the subsequent optimization of the design of seven existing base isolators installed to reduce the seismic risk to antiquities. The Getty Museum's

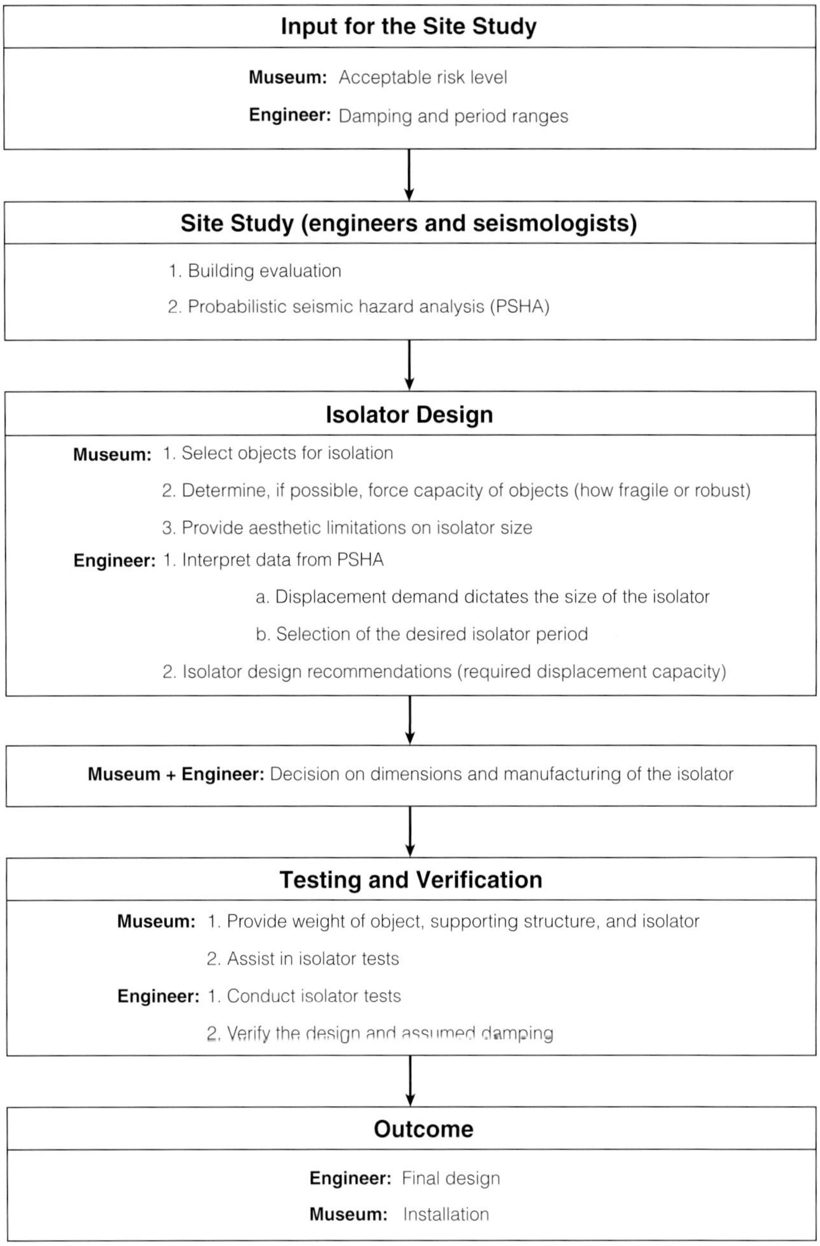

Figure 8
Suggested procedure for base isolator design for museum objects.

base isolation mechanism for individual pieces has evolved over the last twenty years. This study formed part of the Museum's continuing efforts to enhance the seismic protection of its artifacts by incorporating new developments in earthquake engineering. It is strongly suggested that developments in earthquake engineering and up-to-date seismic hazard analyses be implemented in updates of the current design in future studies.

In this study, the seismic design of base isolators was completed for the protection of important works of art. The seismic design of these isolators involved the derivation of dynamic properties, estimation of the seismic demand, and optimization of the isolators' responses. Verification testing provided essential design information and ensured proper isolator behavior. Results of these tests were used in the design of the bilinear spring configuration. The optimized response was obtained through nonlinear dynamic time-history analyses conducted with the design-level ground motions exciting single-degree-of-freedom models of the isolators.

The objective of lowering the force demand due to an earthquake was achieved with the implementation of the base isolation concept. A bilinear elastic isolator response was realized with a stiffness increase during the bilinear portion of the response. This was accomplished by incorporating springs in series to provide the restoring mechanism. In the seismic design, force demands were accounted for, in that the demands were lower than those for an object fixed to the ground, but the demand was not compared to the capacity of each object. A further study of the material strength of antiquities would provide essential information for a more comprehensive design.

References

Abrahamson, N. A., and W. J. Silva
1997 "Empirical Response Spectral Attenuation Relations for Shallow Crustal Earthquakes," *Seismological Research Letters* 68, pp. 94–127.

Chopra, A. K.
2001 *Dynamics of Structures* (Upper Saddle River, N.J.).

Filiatrault, A.
2002 *Elements of Earthquake Engineering and Structural Engineering* (Montreal).

Lindvall, Richter and Associates
1984 "Geologic and Seismic Safety Evaluation of the Main Museum Building for the J. Paul Getty Museum, Malibu, California," Department of Antiquities Conservation, J. Paul Getty Museum, Los Angeles.

Somerville, P. G.
2003 "Magnitude Scaling of the Near Fault Rupture Directivity Pulse," *Physics of the Earth and Planetary Interior* 137, pp. 201–12.

Somerville, P. G., et al.
1997 "Modification of Empirical Strong Ground Motion Attenuation Relations to Include the Amplitude and Duration Effects of Rupture Directivity," *Seismological Research Letters* 68, pp. 199–222.

Stavridis, A., et al.
2006 "Design of a Self-Centering Base Isolator for an Antiquity," *Proceedings of the 100th Anniversary Earthquake Conference, April 18–22, 2006, San Francisco* (Oakland, Calif.).

URS Corporation
2005 "Ground Motion Response Spectra and Time Histories, Getty Museum, Malibu," Department of Antiquities Conservation, J. Paul Getty Museum, Los Angeles.

The Seismic Isolation of Rodin's *The Gates of Hell*

Kimio Kawaguchi

Introduction

Wherever you live in Japan, you are not far from the potential of earthquake damage. Japan sits at the conjunction of several tectonic plates—namely, the Pacific Ocean Plate, the Philippine Sea Plate, the Eurasian Plate, and the North American Plate. As a result, many of Japan's cities are subject to major earthquake damage every few decades.

The Great Hanshin Earthquake, magnitude 6.9 on the Richter scale, struck the Kobe area on January 17, 1995, and generated massive damage. The earthquake destroyed most of the essential services in the modern city of Kobe and in the immediately surrounding area. There was unimaginable damage to art museums, temples, shrines, and historic buildings, as well as to historic districts. This disaster in a modern urban area killed more than sixty-five hundred people, with the majority of casualties caused by the collapse of old buildings as they succumbed to the massive forces of the earthquake. The lack of carefully designed mitigation measures and response plans, as well as their timely implementation, can also be considered one of the major causes of the catastrophe.

Japan's history is full of natural disasters foretelling future threats. In 1923, for example, Tokyo experienced a massive earthquake, the Great Kanto Earthquake, which resulted in more than one hundred forty thousand deaths. Today, simulations of earthquake cycles on a variety of fault lines in the Tokyo area reveal that four major scenarios could possibly be realized in the metropolitan region: an earthquake directly under the city of Tokyo; a second Great Kanto Plain earthquake (in an area 60 km, or 37 mi., around Tokyo); a great Tokai region earthquake (encompassing an area between 100 and 300 km, or about 60–185 mi., from Tokyo); or a major eruption of Mount Fuji (100 km, about 60 mi., from the Tokyo city center). Mount Fuji is classed as an active volcano, the eruption of which would be accompanied by major seismic events. These scenarios will occur in one of two forms: earthquakes with periods of relatively high frequency that occur on fault lines close to the earth's surface, or large-period earthquakes, perhaps generated by distant plates, such as those that have been recorded in Taiwan on September 21, 1999, with a magnitude of 7.6 on the Richter scale, or, in Turkey, the Kocaeli (İzmit) Earthquake, 110 km (68 mi.) east of Istanbul, on August 17, 1999, with a magnitude of 7.4. These earthquakes had periods of as much as 9 or 11 seconds.

The lessons learned from the Great Hanshin Earthquake of 1995 have led to changes in Japan's building code, and today new technologies are being applied to the structural design of buildings and bridges to provide seismic resistance. Some private housing is being built with seismic resistant designs as well. Recently, base isolation technologies have been marketed. Ultimately, however, most of the precautions undertaken have been aimed at earthquakes with high-frequency vibrations and small displacements. Additionally, these new measures for homes and facilities are being realized gradually, in a piecemeal fashion. However, with regard to traditional wooden homes, seismic mitigation measures have not yet been sufficiently considered, and the protection of museum collections is rarely addressed at all.

There remain many unknown factors and inadequacies in developing precautions against seismic events predominantly containing lower frequencies and large displacements, such as those that have occurred around the world (in Taiwan and Turkey, for example). These issues remain paramount in our current seismic mitigation research, planning, and implementation efforts.

Case Study at the National Museum of Western Art

In 1994 the National Museum of Western Art (NMWA), Tokyo, began plans for the construction of an underground exhibition area below the museum's forecourt. In preparation for this construction, Rodin's sculpture *The Gates of Hell,* which was installed in the forecourt area of the museum, was temporarily moved to another site on the museum grounds (fig. 1). The relocation of the sculpture offered the museum a chance to undertake a technical and historical study of the sculpture, as well as to address conservation issues leading to improved stability of its structure and condition. What was not foreseen, however, was the occurrence of the Great Hanshin Earthquake immediately after the sculpture was moved. The tragedy provided important lessons and a new view of seismic hazards for the museum. Rodin's sculpture became a pilot project for the museum's efforts toward seismic protection of its collection.

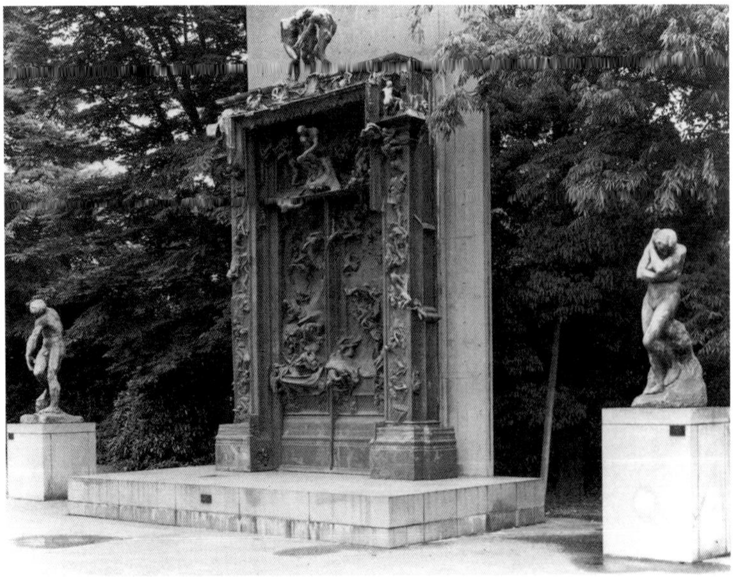

Figure 1
The Gates of Hell installed on its earlier pedestal (around 1960). Photo: NMWA

The Gates of Hell

The traditional view has been that the National Museum of Western Art's version of *The Gates of Hell*, made by the Alexis Rudier Foundry in 1926, was the first casting of the work. This first casting in bronze was not completed during Rodin's lifetime, a fact that has sometimes been seen as a betrayal of the benevolence of Kojiro Matsukata, who around 1926 had presented funds equal to the costs of two casts to the Musée Rodin. However, curators of the Musée Rodin have suggested another theory in recent years that may shed more light on the relationship of Matsukata's patronage of the Rodin Museum and the requisition of material during the Second World War. Reflecting the ongoing stance that the Matsukata Collection (today in Tokyo) was not "returned" to the Japanese but was, rather, a "gift" presented to the Japanese people through the goodwill of the French government, currently published comments by the Musée Rodin state that there was a unique understanding between Matsukata and the Musée Rodin. Further research may clarify the specifics of the history of such a relationship.

The author conducted a historical study and a material, construction, and condition survey of *The Gates of Hell* during the seismic isolation work at the National Museum of Western Art. It has been determined that the NMWA version is most likely the fourth cast of the sculpture, produced around 1947–48, when negotiations were under way between Japan and France regarding the repatriation of items seized as enemy assets by the French government during World War II. The study also determined that it is possible that the NMWA cast and the third cast of the work, today located in Zurich, were both based on the same plaster mold. Unfortunately, the actual order of the casts of this sculpture cannot be clarified until an equivalent and thorough study of the Musée Rodin's cast is undertaken and published.

However, a number of special casting characteristics and a complex rib arrangement of the support structure are visible on the back of the Musée Rodin *Gates of Hell* sculpture. In a careful comparison of these features with those found on the NMWA *Gates of Hell*, the NMWA version shows superior implementation and design of these technical features.

Stabilizing the Sculpture

The NMWA's *Gates of Hell* measures 5.21 m (17 ft.) in height, 3.79 m (12 ft., 5 in.) in width, and 1.03 m (3 ft., 5 in.) in depth. The weight of this massive work is 7.2 tons (after its recent restoration work). The sculpture is made up of twenty-one bronze parts—twenty parts within the main structure, as well as the three figures of Adam in the upper register, which are counted as one part. These parts were originally held together with steel bolts and a steel backing support framework. Each of the bronze parts was created through the sand casting method, and with subtle adjustments for warping and shrinking in the casting process, sophisticated techniques were used to ensure that the individual parts could be perfectly assembled to create a single sculpture. The display base added in 1959 consisted of foundations laid directly into the ground. Back supports for the sculpture and pedestals were then attached to the foundations. The majority of the sculpture was not firmly fixed to the base, however. Of the various sculptures in the collection of the National Museum of Western Art, only *The Gates of Hell* was provided with some form of seismic mitigation, albeit a very simplistic one. Although not necessarily specifically designed for earthquake protection, the concrete backing support and the earthen base did function to stabilize the sculpture to some degree during earthquakes. However, the sheer weight of the sculptural elements themselves meant that the work could never be self-supporting in an upright position. As a result, the work was originally fitted with two box-shaped steel bar supports on the lower sections of both sides, as well as with a central lower rib support. These supports were constructed to bear the majority of the sculpture's weight. However, the outdoor installation of this work led to the corrosion of the lower section of the support structure. Before it was moved, the sculpture and frame had begun to sink, and the sculpture had started to collapse onto itself. A total displacement of 5 mm (about ¼ in.) in height was observed during the last measurement campaign. At the very least, restoration work was needed to replace the structural supports and the bolts of the casting. This restoration, along with seismic mitigation measures, had the overall goal of strengthening the structural integrity of this sculpture and each of its twenty-one parts.

The Musée Rodin's version of *The Gates of Hell* had already had its support framework replaced in the 1980s. However, the conservation report documenting that process was not available when it was requested in 1995.

Seismic Isolation and Reinstallation

After the construction of the underground wing beneath the forecourt, the forecourt was covered with concrete paving. It was decided that, considering the magnitude of possible earthquakes, it would be inappropriate to display the sculpture directly above the underground roof structure of the underground galleries, even if the sculpture were placed on a seismically reinforced base. It was decided that seismic isolation would be developed to lessen the impact of any earthquake forces. The sculpture was reinstalled temporarily from 1994 to 1998 in the forecourt, and budgets were established and plans made for its seismic isolation.

In Japan, seismic isolation fittings are usually developed by general construction companies, and as a result of this protocol, the NMWA sought appropriate contracts to achieve its goal of protecting the sculpture. The following four technical points were included in the design brief:

1. Replacement of the Support Framework and Connecting Bolts The sections of the sculpture are joined into a single unit by 437 bolts that also link the sections to the support structure. Both the bolts and the support structure had to be replaced because of extensive corrosion. Great care was needed in the replacement of the bolts and in the removal and replacement of the ribs in the support structure. New replacement bolts, in sizes M12 and M14, made of SUS304 stainless steel, were used to replace the older, corroded bolts. The new custom-made support ribs were made of SUS316 stainless steel. Given that new stresses could be introduced when sections of the sculpture were rebolted and attached to the sculptural support, and because electrolytic corrosion could reoccur as a result of contact between the different metals (the bronze alloy of the sculpture and the stainless steel of the support), brass washers were used with the bolts, and cast bronze spacers were used at the weight-bearing points of contact between the sculpture and the support framework (figs. 2 and 3).

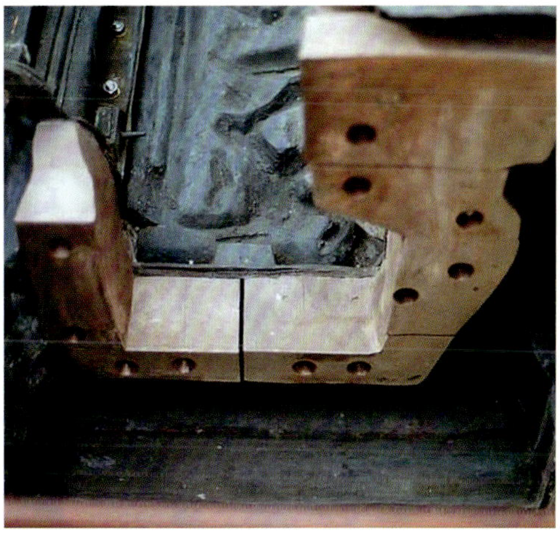

Figure 2
Newly manufactured bronze spacer element installed in the sculpture. Spacers are used to reduce electrolytic corrosion between the different metals used for the sculpture and the new support frame. This particular spacer was divided into five pieces for convenience of installation. Photo: Kimio Kawaguchi.

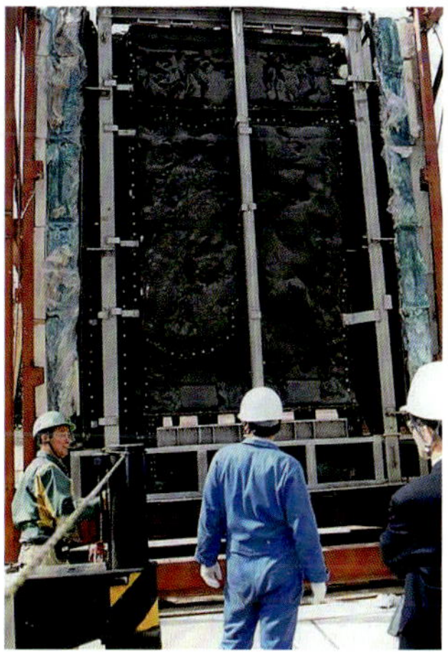

Figure 3
Back view of the sculpture showing the steel support frame joined to the sculptural ribs. Photo: Kimio Kawaguchi.

Bronze spacers were used where the sculpture was joined to the stainless steel support framework. Since the framework supported the majority of the sculpture's weight, the spacers had to be fitted intimately to the form of the sculpture to avoid point loading.

2. Conservation Work on the Sculpture's Front Surface The sculpture had to be placed facedown in order to replace the support rib structure. Sponge rubber was placed around the protruding sections of the sculptural face, and the entire surface was covered with vinyl sheeting. Urethane foam was then sprayed over the entire sheet-covered surface. The hardened urethane foam acted as a protective cushion while the sculpture was resting on its front surface.

3. Design and Construction of a Seismic Isolation Device A seismic isolation pit was constructed to accommodate five seismic isolation devices (fig. 4). Each of these devices was designed to bear 15 tons of the weight of the sculpture and its support structure. The two central devices were fitted with viscous dampers. Once these devices were in position within the pit, a concrete dais was built to hold the sculpture's back support system. *The Gates of Hell* was then fixed to this dais support system. Figure 5 shows the installation of *The Gates of Hell.*

4. Maintenance and Access to the Isolator and the Recording Equipment Space was left between the support structure and the sculpture to provide room for the maintenance of the seismic isolation devices in the pit. The recording equipment placed in this area included an accelerometer, a three-dimensional displacement gauge, and a thermoelectric thermometer. Externally, the sculpture's support backing is fitted with an anemometer, a global positioning system (GPS) device, and a wind direction gauge. These devices provide

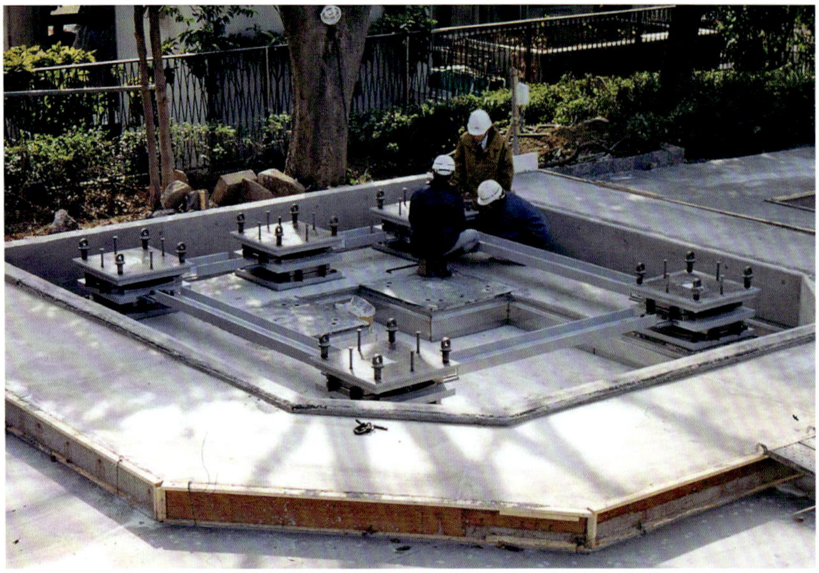

Figure 4
Isolation pit after the installation of the five isolator units, prior to the installation of the sculpture. Photo: Kimio Kawaguchi.

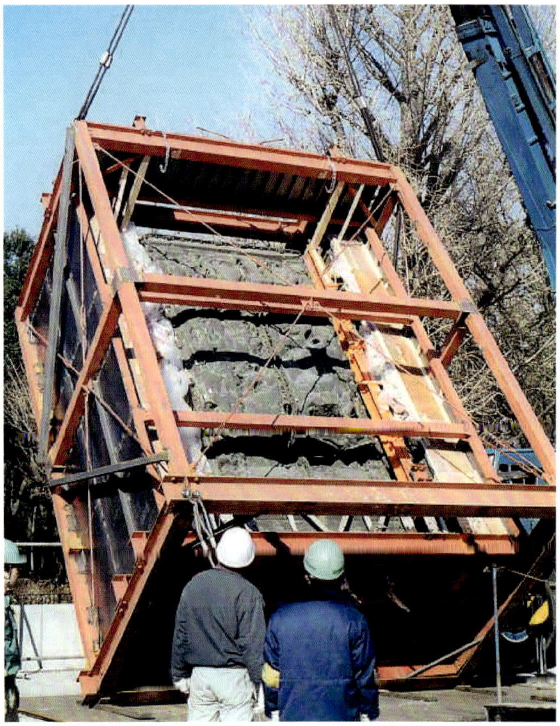

Figure 5
Installation of *The Gates of Hell* in 1999. Photo: Kimio Kawaguchi.

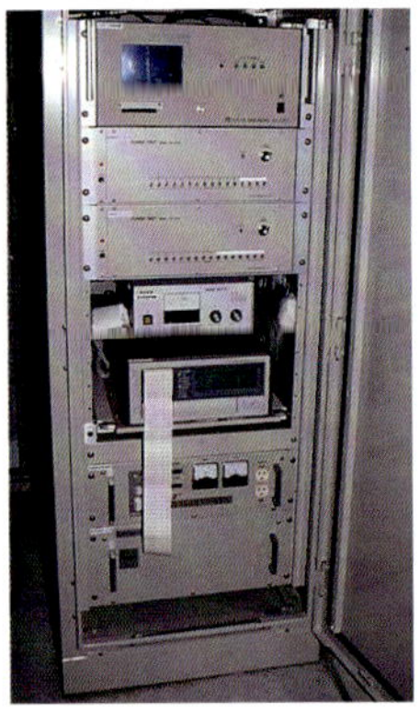

Figure 6
Monitors that capture data from recording devices that are attached to the interior surfaces of the sculpture and support structure. Photo: Kimio Kawaguchi.

records of the forces affecting the sculpture when the seismic isolation devices are triggered by either seismic activity or a typhoon (fig. 6).

Seismic Isolation Devices

Seismic isolation devices, using a number of basic design principles, are made in Japan. A typical isolation device combines a rolling element, such as a roller bearing, ball bearing, or wheel, with a slipping device, or low-friction surface such as Teflon, which allows the base to move freely in any direction.

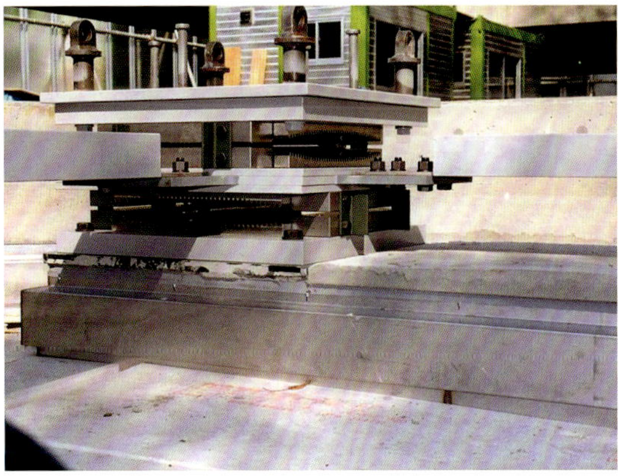

Figure 7
An individual vertical roller bearing. A steel structure attaches all the isolators together. Two of the devices are located near the back of the sculpture, in order to support the majority of the weight of the sculpture and support structure. Photo: Kimio Kawaguchi.

Circular roller bearings were chosen as the moving parts for *The Gates of Hell* isolator, with the rollers placed between upper and lower curved elements (fig. 7). Pinions were attached to each end of the rollers in order to prevent twisting. Two layers of these roller assemblies were set at right angles to each other, thus allowing 360 degree movement at the top level of the isolation base. Given the low coefficient of friction achieved by this device, three adults were able to laterally displace the 60 ton sculpture and base unit with relative ease. The ease with which the sculpture assembly could be displaced on the isolator raised a concern that even slight winds might set such a structure into motion. The upper and lower elements were designed with a semicircular radius arc of 200 cm (78¾ in.). The characteristic vibration period of the assembly was designed to be 4 sec., in order to limit the response acceleration to 100 Gal (0.1 g). The range of movement of the rollers was designed to be 22 cm (8⅝ in.), in order to create a total response displacement of 40 cm (15¾ in.). Concern was also raised about the amount of time the structure might continue to move during a seismic event. Because of these concerns, the structure was fitted with viscous dampers to absorb some of the excess energy. The dampers assure that the device will not move until it is subject to a 30 Gal (0.03 g) lateral force.

To evaluate the isolator, a seismic force of 50 kine (equal to a velocity of 50 cm/sec., or 1.6 ft./sec. [1 mph]) was applied to the construction.

Acceleration Levels Recorded in Recent Strong Winds

Given that Japan is prone both to earthquakes and to typhoons, the sculpture also had to be set to be protected in winds up to 60 m/sec. (197 ft./sec., 134 mph). Ultimately these devices were designed to ensure the sculpture's safety under simultaneous forces of 100 Gal (0.1 g) and a base wind speed of 38 m/sec. (124.67 ft./sec., 84.9 mph).

Strong wind can cause motion in this seismic isolation system because of the configuration of the sculpture, with its large exposed flat surface, particularly if the wind is blowing normal to the broadest dimension of the sculpture. The system is designed to not exceed maximum displacement and to maintain response acceleration below 100 Gal (0.1 g), even in the presence of the following two conditions:

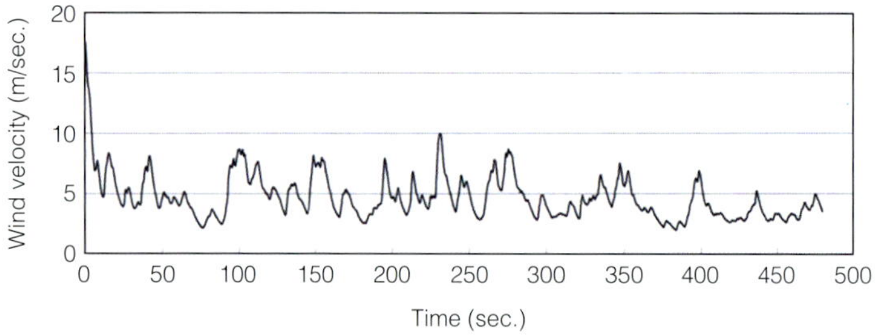

Figure 8
Wind velocity at the sculpture site in a strong wind. Winds occasionally exceed 15 m/sec. (49.2 ft./sec., 33.5 mph), but their individual duration is brief.

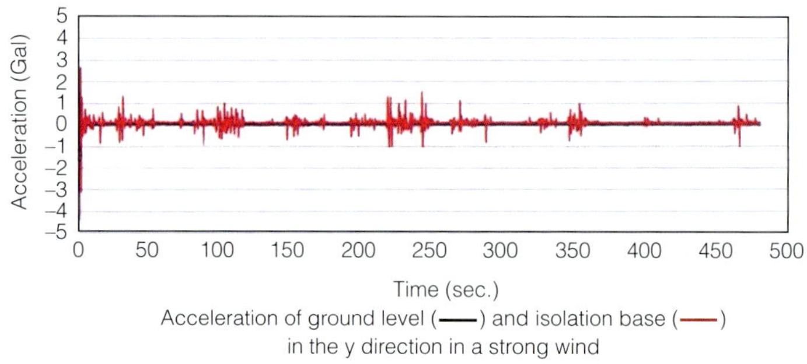

Figure 9
Response of the isolator during a strong wind. The maximum wind velocity measured was about 20 m/sec. (65.6 ft./sec., 44.7 mph), and in that instance, the resulting acceleration of the seismic isolation system was less than 5 Gal (0.005 g) in the y direction. Thus, only a small force affected the sculpture during that incident.

- In continuous wind of 60 m/sec. (196.85 ft./sec., 134 mph), with an annual probability of 1/32,000
- When 38 m/sec. (124.67 ft./sec., 84.9 mph) winds occur simultaneously with 100 Gal (0.1 g) seismic force, with an annual probability of 1/100

The data shown in figure 8 consist of velocity measurements taken during a recent strong wind event. The measurement trigger for the device is set to 15 m/sec. (49.2 ft./sec.). When wind velocity exceeds this trigger load, data logging starts and continues for 480 sec. at 12.5 measurements per second, resulting in 6000 data entries. Figure 9 shows the response of the sculpture on the isolator during a strong wind.

Calibration of the Base Isolator

While it is obvious that no single mechanical device can withstand all types of earthquakes, the device was calibrated to withstand Great Hanshin (JMA Kobe), Hachinohe, El Centro, and Taft earthquake forces. However, it is thought that this device would not work well in the instance of a long-wave earthquake, such as the type experienced in Turkey or Taiwan. If the characteristic vibration period of the earthquake is longer than 4 sec. and the response displacement is within 40 cm (about 15½ in.), then the viscous dampers first decrease the seismic movement. If further seismic movement occurs beyond 40 cm, the sculpture's pedestal would break. Since there is no vertical isolation if these limits are exceeded, then the unit would uplift, although the breadth of the base would probably prevent it from overturning, should it begin to rock or deviate from true vertical.

Viscous Dampers

A high-molecular-weight, butane-type hydrocarbon compound with high viscosity (SA-P), developed by Oiles Corporation, was used for the viscous

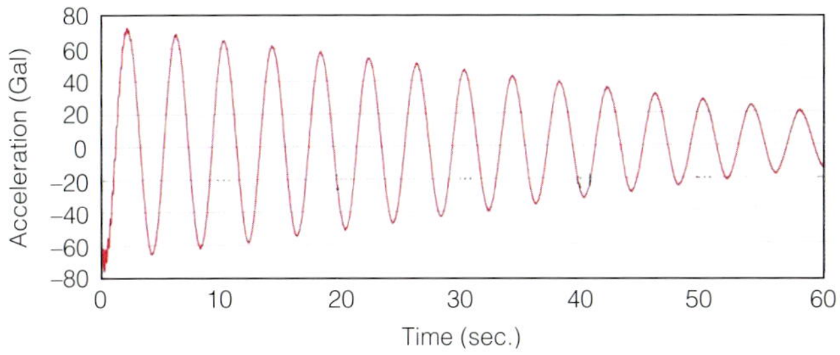

Figure 10
Acceleration of the base in the y direction during a free vibration test without dampers. Tests conducted without the dampers produced a natural period of the circular roller bearing that met design values (3.99 sec.).

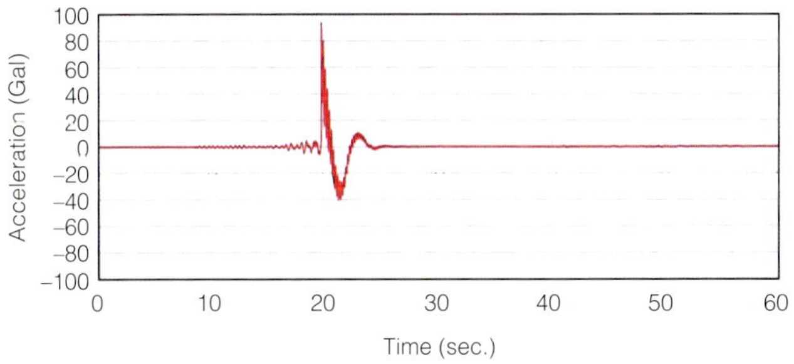

Figure 11
Acceleration of the base in the y direction during a free vibration test with dampers. Tests revealed that the viscous dampers slowed the movement of the isolation base. Movement stopped in one cycle, and the residual displacement was negligible.

dampers. This material was placed in shallow containers, and cylindrical steel plates were allowed to move horizontally in the viscous compound, thus lessening seismic force. Free vibration tests were conducted to confirm the efficacy of these dampers and their ability to provide a returning force for the isolator. In the tests, an oil jack pulled the isolation base horizontally. When the base was released from its displaced position, measurements were taken of the base's horizontal and vertical acceleration and of its horizontal displacement as movement continued. Results are shown in figures 10 and 11.

These results indicate that, as per design, the system is capable of reducing seismic forces acting on the sculpture. However, the dampers can also reduce the effectiveness of the isolator under certain conditions, such as when the earthquake movement is of very low frequency, when displacements are great, or when the surrounding temperature is low (since the damping effect is somewhat temperature dependent). In such cases the damper resistance reduces the system acceleration less than expected.

Measurement devices were attached inside *The Gates of Hell* after construction was completed, and they are used to measure daily temperature and humidity levels, wind speed, and temporary deformation of the sculpture due to earthquake or wind loads. Devices have also been placed in five additional locations: in the underground gallery interior, in the seismic isolation pit, at the level of the sculpture's base, in the middle of the sculpture, and at the top of the sculpture. This group of devices is designed to measure the relationship of the acceleration of the sculpture to the acceleration of the surrounding area and structures.

In July 2005, sensors recorded movement data for *The Gates of Hell* during the North Chiba Earthquake (fig. 12). All movements recorded were in agreement with those created in test simulations (fig. 11).

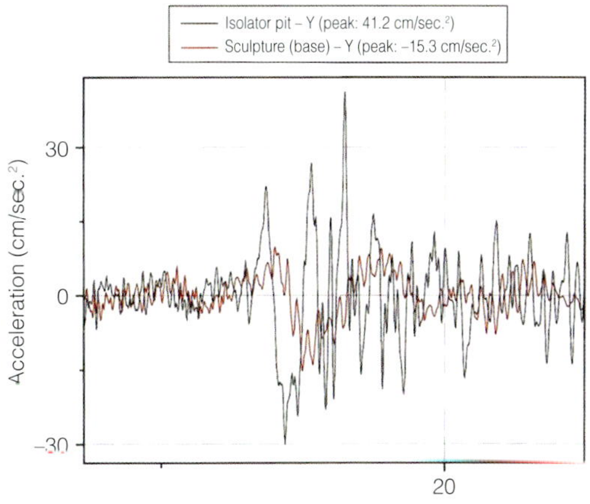

Figure 12
Acceleration of the isolator pit and the sculpture base measured in the y direction on July 23, 2005, during the North Chiba Earthquake (magnitude 6 on the Richter scale), which struck approximately 35 km (21.7 mi.) from Tokyo. The black line shows the acceleration of the isolator pit, and the red line shows the acceleration of the sculpture. The peak area shows that the maximum force on the pit was 41.2 Gal (0.04 g), while the maximum force at the sculpture's base was 15.3 Gal (0.01 g), confirming that the isolator responded within design limits.

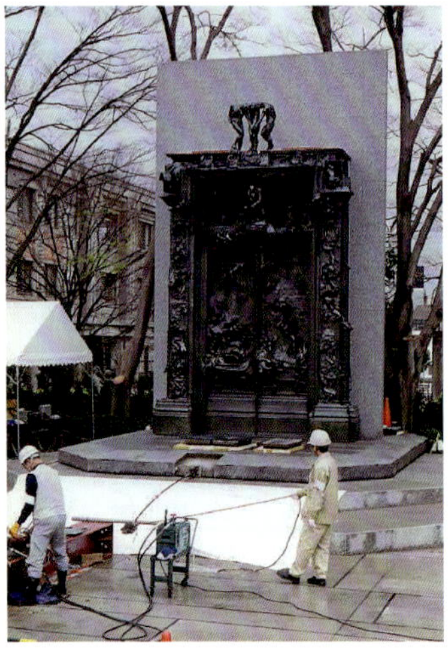

Figure 13
The Gates of Hell on its newly installed base isolator and pedestal. Photo: Kimio Kawaguchi.

Conclusion

The approach taken toward this project offers one solution to the issue of seismic mitigation for *The Gates of Hell* (fig. 13), but it is by no means a complete answer. It is unlikely that there is a way to fully protect the sculpture from all seismic activity. The installation was planned to address the seismic damage that could be conceived, given the geographic and geological location of the museum. Public access issues also guided the approach, as did the aesthetic dictates of museum display.

Disaster preparedness in Japan's art museums must strive to develop new approaches to seismic mitigation for the future. However, most art museums in Japan do not have conservation departments and are lacking in seismic mitigation measures. Additionally, disaster planning and preparation are not defined as part of the duties of curatorial staffs, a fact that weakens the possibility that such programs will be developed and implemented.

The National Museum of Western Art understands the importance of disaster planning as part of its responsibilities as a national art museum. We hope to provide pioneering work and thus encourage other art museums to follow our lead. In addition to the work on *The Gates of Hell*, we have conducted experiments using different types of seismic isolation devices under other sculptures in the museum's forecourt. We are currently studying ways to prevent pedestals from overturning when the seismic force is approximately 280 Gal (0.28 g).

To make such efforts economically feasible, we have begun new experiments that include attaching a Teflon-coated stainless steel slip plate to the undersides of sculpture pedestals in order to prevent tipping or overturning. The plate widens the footprint of the pedestal-sculpture assembly and positively affects the base-to-height ratio toward stability. The plate also reduces the coefficient of friction between the floor and the pedestal, making sliding more likely and making overturning due to friction less likely.

Research in seismic mitigation can take numerous directions, especially if there are no budgetary constraints. However, we hope that our future efforts will result in the development of low-cost seismic mitigation measures, making them more affordable to a larger number of museums.

Seismic Risk Mitigation in Istanbul Museums

Mustafa Erdik **Eser Durukal**

Nevra Erturk

Introduction

Istanbul is continuously threatened by earthquakes that endanger lives and the preservation of its material heritage. Historical records of Istanbul over the last two thousand years indicate that the city experiences a medium-intensity earthquake (i.e., VII–VIII on the Modified Mercalli scale) every fifty years and a high-intensity earthquake (i.e., VIII–IX) approximately every three hundred years. Studies of both the historical and contemporary seismicity of the region, as well as geophysical and geological research, support the expectation of a major earthquake, of a magnitude larger than VII on the Modified Mercalli scale, in the very near future. Such an earthquake is likely to originate from the fault system of the Marmara Sea region. These faults are approximately 20 km (12 mi.) south of Istanbul. Given the proximity of these major faults and the magnitude of the earthquake expected, it is clear that the museums and cultural collections of Istanbul face great threats.

It is a well-known fact that throughout history earthquakes have caused significant damage to historical structures in Istanbul. The most recent damage was suffered as a result of the 1999 Kocaeli Earthquake, some 80 km (about 50 mi.) away from the city. The museums and libraries affected by this earthquake included the Museum of Turkish and Islamic Arts, the Topkapi Palace Museum, Beyazit Library, Suleymaniye Library, the Dolmabahce Palace Museum, the Ragip Pasa Library, the annex building of the Istanbul Archaeology Museums, and the Istanbul Painting and Sculpture Museum.

There have been several attempts to draw attention to the earthquake vulnerabilities of historical structures in the city. While these efforts have taken the form of conferences, workshops, and project proposals (some of which have been implemented), less is known and developed with respect to the safety and protection of museum collections. Earthquakes threaten not only the museum buildings but also the artifacts on display and all objects kept in storage. This threat is of particular concern since there are about fifty museums in Istanbul, some housing world-renowned collections.

Additionally there is a threat to all of the peripheral equipment, such as the display and storage furniture, the electrical and electronic equipment, heating, air-conditioning, and lighting systems. These are also important aspects of the museum environment and must be protected, since failure of any one of them may result in eventual damage to the objects. Although some awareness about the earthquake vulnerability of museums is beginning to surface and is beginning to be discussed, few advances have been made toward development and implementation of solutions and mitigation schemes for collections. Many of the buildings serving as museums in Istanbul are themselves of historic importance. Most are rigid masonry structures and were built for purposes other than the housing of collections, making them inherently inadequate as museum buildings. Even if we assume that such buildings will withstand the effects of expected earthquakes, the high floor accelerations that would result in such rigid buildings will surely cause heavy damage to contents, which in this case are the precious objects being displayed and stored.

If we look beyond the structural concerns related to the building itself, we find that in many museums, the objects themselves are not properly displayed to ensure stability during an a seismic event. The storage areas are often overloaded, with no seismic mitigation precautions in place at all. This is particularly unfortunate since storage methods do not necessarily need to be constrained by the same aesthetic issues that may challenge effective seismic mitigation efforts in public exhibition spaces. In storage areas, seismic mitigation efforts can be much more straightforward.

Considering the number of endangered museums in Istanbul and the size of their collections, assessing the earthquake risk and then mitigating that risk are immense tasks by any standard.

This paper provides a status report on the current earthquake risk mitigation efforts by the museums in Turkey. It summarizes several projects that have been carried out over the past few years to protect museum collections from earthquake damage, and it defines future actions that would mitigate earthquake risks associated with museum buildings and their contents.

Mitigating Damage from Earthquake-Induced Ground Motion

Efforts to address the protection of museum objects were begun with the study of the dynamic responses of rigid bodies on horizontal planes. In this connection, three elementary forms of response were identified: sticking, rocking, or sliding. The overall response of an object to seismic motion can generally be defined by some combination of these three specific response categories.

The pioneering work conducted by Agbabian and Augusti and their colleagues constitutes some of the first published scientific studies on the seismic response of museum objects and has established the basic guidelines to be followed in designing seismic resistant mount supports (Agbabian, Masri, and Nigbor 1990; Agbabian et al. 1988; Augusti, Airoldi, and Ciampoli 1992).

It is a given that excessive, unrestrained sliding or rocking must be avoided, since it may result in an object falling from its support or colliding with other objects, with display or storage furniture, or with walls.

Nevertheless, controlled sliding can reduce stresses due to inertial forces and thus be beneficial. Controlled sliding is difficult to achieve, however, since it depends upon the particularities of a given three-dimensional ground motion induced by a specific earthquake. The dynamic variability of the coefficient of friction between the adjacent surfaces also complicates the response. The coefficient of friction can be accurately defined. And when a low enough coefficient of friction between the object and the support plane is chosen, the range of objects that slide rather than rock when subjected to earthquake forces can be extended. The accurate prediction of ground motion, however, remains elusive and can, at best, consist of an educated guess based on historical seismic studies, building studies, geotechnical studies, and probabilities. This fact is particularly unfortunate for mitigation efforts, since the period and acceleration are key factors in earthquake damage.

Once the ground acceleration reaches a certain value, the rocking of the object will likely begin. This value is defined by the aspect ratio of the object (modeled as a rigid body)—that is, the ratio of the height of the object to the width of its base. The dynamic motion of the object can then be defined as a series of impacts between the object and the plane of support. Although these series of rocking impacts can dissipate energy and ultimately prevent overturning, they can also cause direct damage to the object. For the body to overturn, it must be subjected to a sufficiently large ground motion velocity. Agbabian, Masri, and Nigbor (1990) have defined an equivalent height for the assessment of an object's rocking susceptibility.

Pedestals and showcases of certain unfavorable dimensional aspect ratios can increase the intensity of ground motion and modify its frequency. However, the dynamic response can be improved by changing the display furniture's structure, dimensions, and support conditions. In the latter instance, this can be achieved by anchoring the furniture to the building or allowing these items to slide easily on the floor. This is, however, generally decided on a case-by-case basis and can, as a result, be quite expensive, since it depends on the location of the furniture within the building or room. It is well known that different parts of the building (for example, the upper versus lower floors, or the center of a large room versus the areas near the structural walls) respond differently to earthquake forces. The effectiveness of this approach is also influenced by the type and number of objects housed in showcases or supported by pedestals.

In general, the earthquake protection of museum objects encompasses the following provisions:

Strength For objects on fixed supports (especially if they are slender), the stresses generated by the earthquake motion (i.e., the motion of the support) need to be assessed. To reduce these stresses, the object can be strengthened (with minimum modification of the object's physical, mechanical, and chemical properties), or the object can be allowed to slide in a controlled manner, or it can be base isolated.

Rocking and Overturning For freely supported museum objects with aspect ratios (or slenderness) prone to rocking and overturning, the remedial protective measures can be: fixing the object to the support plane (bolting it to the floor or wall), lowering its center of mass (adding weight to the lower section), enlarging the base (to favorably influence the aspect ratio), and/or allowing for controlled sliding.

Sliding It is important to assess as accurately as possible the sliding displacement trajectory of the object in order to prevent the object from colliding with other objects and/or the walls of the room and/or the glass sides of a showcase—or, indeed, to prevent the object from falling from its shelf or pedestal.

Base Isolation For protection against earthquake risks, the implementation of seismic base isolation has been of practical interest only for a limited number of collections, and even then, only for a few types of objects with high importance and fragility. This is predominantly because of the high development and production costs involved.

These costs, however, can be somewhat reduced overall by reducing the number of isolators and by grouping the objects. Most objects in small museums can be treated as groups of rigid bodies and are usually exhibited in special showcases as groups. Although individual support systems for such objects can be made, it would be more efficient to utilize base isolation approaches for entire cases. The same principle holds true for racks, cabinets, and cases in storage areas.

Museum Risk Mitigation Projects in Turkey

Following some embryonic studies supported by UNESCO in 1992, a report and a brief manual on earthquake protection of objects displayed in museums was prepared by the Department of Earthquake Engineering of Bogazici University. Additionally, limited applications of these principles have been undertaken at the Topkapi Palace Museum in Istanbul.

In 2004 a World Bank ProVention project entitled "Seismic Conservation of Historical and Cultural Treasures of a World City: Sizing the Need and Formulating an Action Plan for the Museums of Istanbul, Turkey" was carried out by a team consisting of Bilgen Sungay and Nevra Erturk under the supervision of Marla Petal (Sungay and Erturk 2004). The goal of this project was to enhance knowledge regarding disaster preparedness, focusing particularly on nonstructural mitigation, in order to save lives and prevent injuries to museum staff and visitors. The effort also sought to preserve our cultural heritage for future generations and to protect business continuity in the tourism sector, as well as to assist this sector in prioritizing and developing practical nonstructural seismic mitigation action plans. The project has developed forms for the assessment of risk, compiled examples of hazards and best practices, and prepared educational materials for nonstructural hazards and mitigation methods for museum collections. The forms prepared include a rapid room survey form (for classification of objects that need to be protected by either simple, object-specific methods or by sophisticated mitigation methods), a rapid survey summary form (for synthesis of the results of the first form to portray the museum-wide results and determine the approximate cost of the mitigation), and an object risk identification form (for making decisions on the application of object-specific and sophisticated mitigation methods).

The project has identified the following earthquake risks to the displayed objects in museums:

- Risks from broken glass (window, door, skylight, or showcase glass)
- Risks caused by impact—that is to say, when items fall or collide
- Risks within showcases (when objects overturn, slide, and/or collide with other objects; when lighting fixtures or glass shelves collapse)
- Risks to hanging objects (swinging and collision, slippage, or support/restraint failure)

- Risks from the ceiling or from ceiling-mounted fixtures (falling suspended ceilings, light fixtures, ducting, signs, and/or pipes)

The project has suggested the use of the following applications for the earthquake protection of museum displays:

- Cautious and appropriate use of small amounts of wax, monofilament, and/or plastic-encased steel wire to secure objects
- Use of metal hooks with an appropriate failure strength to provide an adequate safety factor for suspended objects, in order to reduce the risk of falling
- Bolting or screwing objects to their support planes to reduce the risk of toppling
- Securing objects using specially produced mounts (such as contour mounts)
- Placing sandbags inside the objects to lower the center of mass and thus reduce the risk of toppling
- Placing rubberized shelf mats under small objects to reduce the risk of sliding and to minimize damage by impact if overturning occurs (one should note, however, that the object must be somehow supported or restrained to prevent overturning)
- Using latches to prevent showcase and cabinet doors from opening
- Placing restraints on open shelves to reduce the risk of objects falling off the shelves
- Using padding between objects (especially in storage areas) to minimize collision damage and abrasion
- Covering glass surfaces with security film to prevent shattering
- Designing new shelving systems with improved structural integrity
- Base isolation applications

In 2006 a second World Bank ProVention project, entitled "Protection of Museum Items against Earthquake Shaking by Low-Cost Base-Isolation Devices," was conducted by Gokturk Onem and Cuneyt Tuzun under the supervision of Mustafa Erdik and Eser Durukal (Onem et al. 2006). The project's aim was to develop inexpensive ball-in-cone type isolation units (BNC devices) in standardized geometries and load capacities which could easily be used by museums for the earthquake protection of displayed and stored objects (fig. 1).

For several museums, housing vast amounts of cultural and historic artifacts is problematic. In comparison to displayed items, objects kept in storage might be even more prone to earthquake damage because of their high concentration and instable storing conditions—conditions that are, in many cases, caused by the lack of knowledge of safe storage techniques. And, since objects with different geometric configurations and weights are often displayed or stored in the same showcases or racks, the solutions should be all encompassing, easy to implement, and reasonably priced.

Using BNC isolators under showcases is both an economically and technically viable approach. Isolating a whole display or storage floor or floor

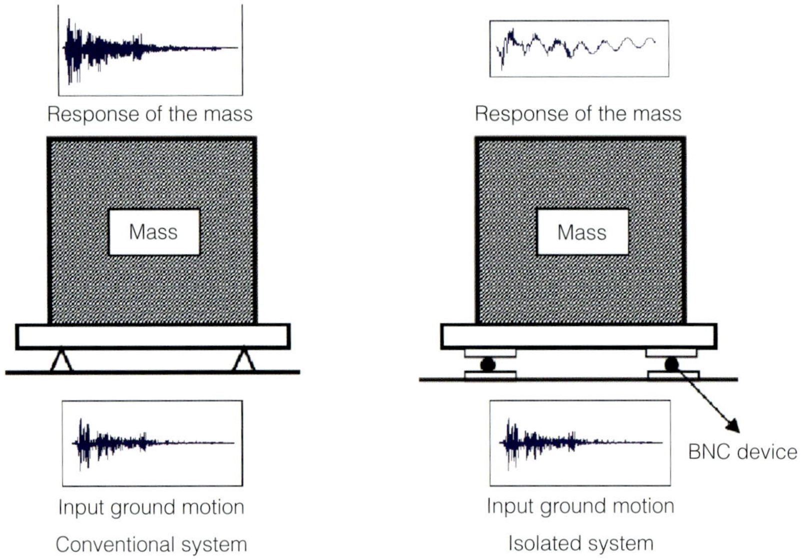

Figure 1
Comparison of conventional and base isolation systems for earthquake protection of museum displays (BNC = ball-in-cone).

area through the use of multiple isolators might constitute an even more holistic solution. Figure 2 illustrates these isolator units, which can be used for showcase or floor isolation.

The ball-in-cone type isolation units were designed with the aid of analytical techniques and computer simulations. A prototype system was then produced, and the experimental response of this system was verified on an earthquake shake table using replicas of museum display and storage furniture and representative models of museum objects. The project found that the use of ball bearing isolation systems for the protection of museum items was an effective high-performance, low-cost, and easy-to-implement solution for reducing the seismic risk.

Earthquake Protection Measures in Istanbul Museums

As mentioned above, many of the collections in Istanbul have not been properly protected against strong earthquake forces, and as a result, many objects are highly susceptible to damage due to rocking, overturning, or collision (fig. 3). There is, however, an increase in awareness, and attention and

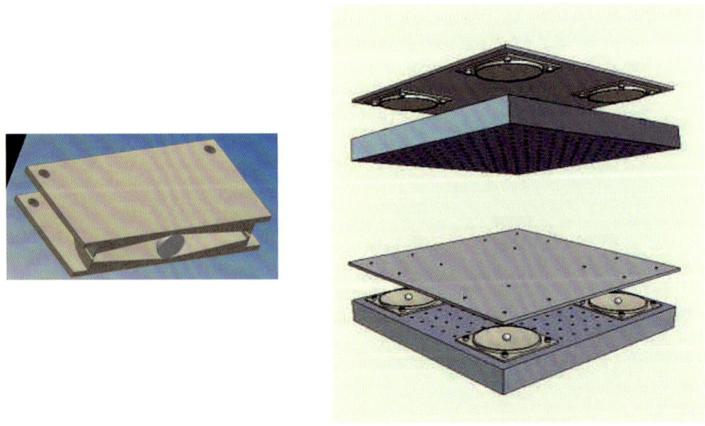

Figure 2
Views of isolating platforms, showing the use of BNC devices.

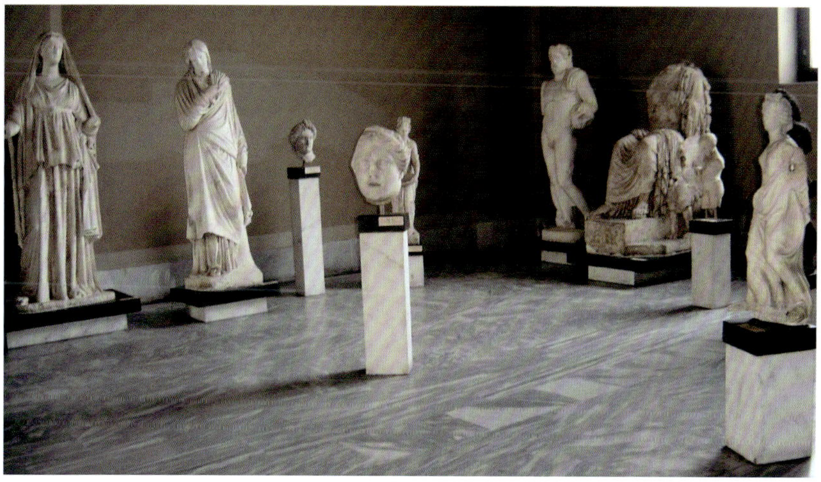

Figure 3
Objects on display in the Istanbul Archaeology Museums.

resources are being devoted to studying and developing approaches to risk mitigation in museums. The institutions involved in these efforts include the Turkish Ministry of Culture and Tourism General Directorate of Cultural Assets and Museums; the Kandilli Observatory and Earthquake Research Institute and Department of Earthquake Engineering Disaster Preparedness Education Program, Bogazici University; and the Museum Studies Graduate Program of the Faculty of Art and Design, Yildiz Technical University. Through the efforts of these institutions, assisted by the visits of Jerry Podany, Head of Antiquities Conservation at the J. Paul Getty Museum, and the associated transfer of information and technology (Podany 1991), the following museums have implemented various protection measures for the mitigation of earthquake risk to displayed and stored objects: the Topkapi Palace Museum, the Sadberk Hanim Museum, the Istanbul Archaeology Museums, the Ancient Orient Museum, and the Rahmi M. Koc Museum.

Although some awareness about the earthquake risks that the museums are exposed to has been raised, it can be said that at this point little implementation has been realized. When considering initiating seismic damage

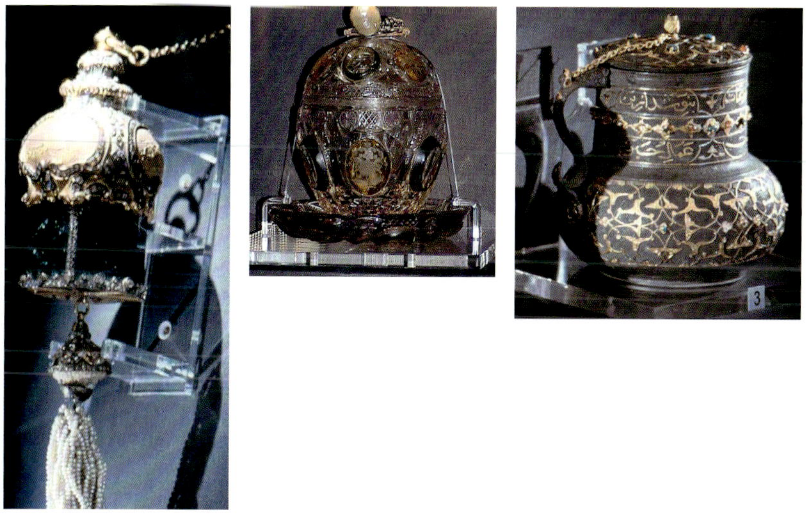

Figure 4
Installations using custom mounts and monofilament in the Treasury Section of the Topkapi Palace Museum.

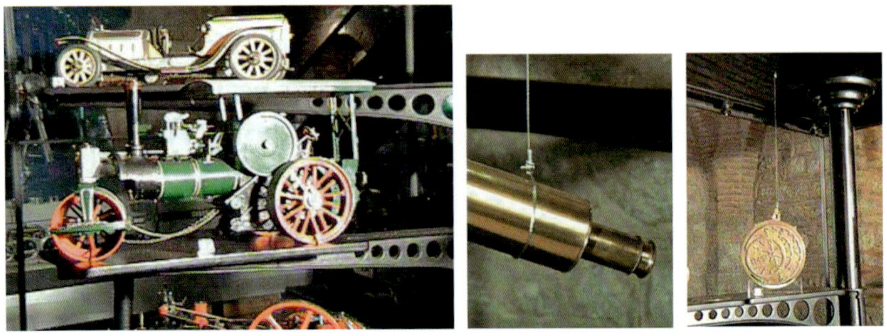

Figure 5
Installations using custom mounts and monofilament in the exhibition galleries of the Rahmi M. Koc Museum.

mitigation, museums face many challenges, which will vary depending upon the museum and its administrative status, its physical conditions, the earthquake awareness of its staff, and, of course, the extent and nature of the collection. The many challenges include:

- The need for special permission to fasten objects to the walls, ceilings, or floor in museum buildings of historical importance. Anything that might alter or damage the structure is difficult to defend and undertake.
- Frequent changes of management and of general administrative staff, and changes in the overall priorities set by these staff members.
- Limited budget that can be devoted to risk mitigation.
- Insufficient number of staff and limited time available from the existing staff to address damage mitigation programs.
- Lack of a well-designed volunteer system (this applies to most of the museums).
- A general reluctance to undertake mitigation measures because they might change the object or because the process will create long-term maintenance problems that the museum is not able to resolve.
- The greater priority placed on addressing the obvious necessity of structurally strengthening and improving the museum building; once this process is complete, the collection is often assumed to be safe within a retrofitted building.

Issues Related to Earthquake Retrofitting of Museum Buildings

Although in the past the focus of earthquake protection of vulnerable buildings has been toward the improvement of the load bearing structure, performance based engineering has, in recent times, brought more attention to the protection of nonstructural components and contents—an interest that has developed in parallel to structural efforts. Increasing the strength and stiffness of the structure (by, for example, adding structural walls and jacketing columns) has generally been adopted as a conventional method for improving the structural capacity of a building. Such a procedure can also protect some of the nonstructural elements through a decrease in the relative floor deformations; however, it is generally detrimental to the contents of the buildings, since it increases the floor response. This problem is especially a concern for buildings

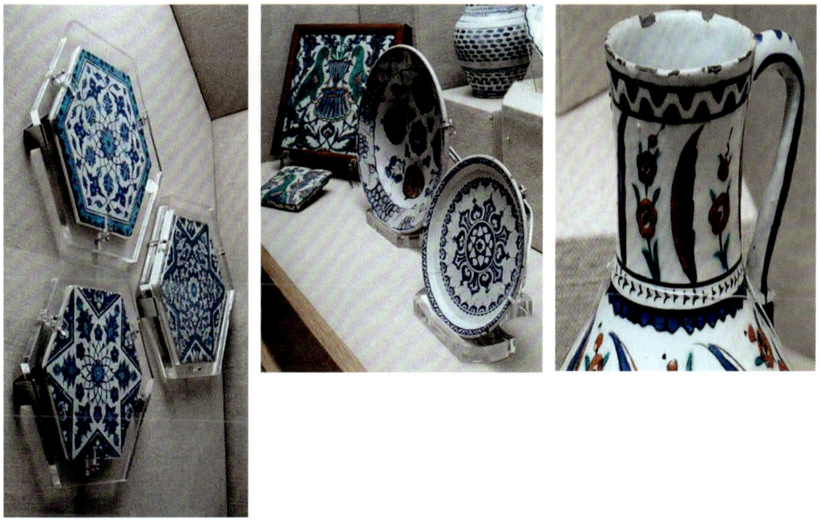

Figure 6
Installations using custom mounts and monofilament in the exhibition galleries of the Sadberk Hanim Museum.

housing expensive and critical or irreplaceable contents, such as hospitals and museums. Seismic base isolation of the whole building is often seen as the general solution for earthquake protection (structural, nonstructural, and contents). However, broad application of this approach has yet to take hold.

Suggestions for Comprehensive Earthquake Risk Mitigation in Museums

There is generally a problem with limited budgets. Even so, it is possible to start instituting inexpensive and practical nonstructural mitigation methods immediately. This initial action would contribute to a greater focus on the more complex methods later. It is important to develop interdisciplinary consulting teams that can participate in the initial efforts of nonstructural mitigation across museums.

Earthquake risk mitigation should be done not only in the exhibition galleries and storage areas but also in offices, museum shops, exits, corridors, halls, and other public service spaces.

A comprehensive program for the mitigation of earthquake risk to museums should start with the preparation of an earthquake loss scenario specific to the site, building, and collection being addressed. This scenario can be used for sensitizing the decision makers, for raising the awareness of professionals, and for prioritizing actions to safeguard the museum buildings and

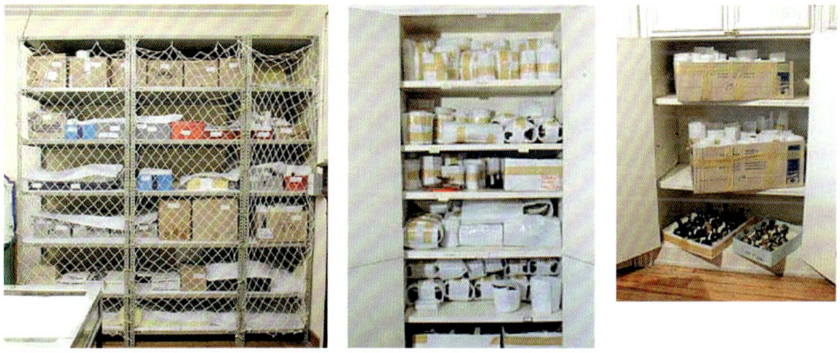

Figure 7
Storage areas of the Sadberk Hanim Museum.

their contents. Thus it establishes a basis for vulnerability reduction plans and decisions on behalf of policy makers. Earthquake loss assessment should be expressed in terms of damage to a museum building and its contents under a given scenario of earthquake ground motion. The earthquake risk assessment will focus on two particular areas of concern: the museum buildings and the objects that the buildings house. In the first stage, the damage potential of the museum buildings should be assessed on the basis of the expected site-specific earthquake ground motion scenario, with the structural, architectural, and historical characteristics of the building taken into account. The second stage will consist of the same assessment carried out for the objects. Given the immense size and diversity of most collections, it will be necessary to group the existing inventory (objects of different size, material, state of vulnerability, and so on) into typical classes. The outcome should be the earthquake vulnerability assessment of the museum buildings combined with the vulnerability assessment of the museum contents.

The following activities should be conducted for the earthquake vulnerability assessment of museum buildings:

- Earthquake hazard assessment: site-dependent deterministic assessment of the expected earthquake scenario and ground motion at the site of the museum building(s).
- An assessment of the structural properties, architectural characteristics, and material properties of the various museum buildings.
- An assessment of the earthquake vulnerability of these buildings: determination of the response of different structural types in the inventory and assessment of their damageability.

The following activities should be conducted for the earthquake vulnerability assessment of museum contents:

- Inventory assessment: a thorough characterization of the types and characteristics of the objects on display and in storage, including: a determination of general types of objects in terms of size, material, and so on; an assessment of acceleration- and drift-sensitive objects; and a general estimation of story-level ground motions.
- Shake table tests: These tests—conducted on models or object replicas—should be done on the object types for the purpose of determining their response to expected earthquake ground motions.
- Determination of the critical level of ground motion that would create instability of the various categories of objects during earthquakes; the level of ground motion required to topple/damage each object type should be assessed.

The following criteria will be used in evaluating loss assessment and putting into place recommendations for structural and nonstructural mitigation of the museum building and contents:

- If a museum building is expected be subject to heavy damage or collapse, it will be assumed that its contents will be predominantly lost.

- If a museum building is expected to suffer midrange to no damage, then the emphasis will be placed on the risk posed to the displayed objects (as well as objects in storage), with the assumption of and further recommendation for structural retrofit.

The project should provide an estimate of expected losses in museums and guidance regarding prioritized treatment to address expected damage to the museum building (structural and nonstructural), displayed objects, and stored objects. Activities to raise awareness and to create a basis for sponsorship should also constitute an important part of this study. Public relations, advocacy, and promotion (e.g., identification of sponsors and advocates) for mitigation should be an integral part of the activities. The project report should summarize all steps of the study, evaluate and discuss the outcomes, and comment on possible implications of the results; this report should be submitted to appropriate authorities to serve as a master plan for risk mitigation in museums.

For future applications, it is very important to design new museums according to proper earthquake performance criteria for both structural and nonstructural components through the use of base isolation and other structural control techniques. When this procedure is followed, damage mitigation will become part of building projects at their inception, and the collections within our museums and the cultural heritage they represent will be safe from earthquake damage.

References

Agbabian, M. S., S. F. Masri, and R. L. Nigbor
1990 *Evaluation of Seismic Mitigation Measures for Art Objects* (Los Angeles).

Agbabian, M. S., S. F. Masri, R. L. Nigbor, and W. S. Ginell
1988 "Seismic Damage Mitigation Concepts for Art Objects in Museums," *Proceedings: Ninth World Conference on Earthquake Engineering* (Tokyo).

Augusti, G., L. Airoldi, and M. Ciampoli
1992 "Mitigation of Seismic Risk for Museum Contents: An Introductory Investigation," *Proceedings: Tenth World Conference on Earthquake Engineering* (Rotterdam), pp. 5995–6000.

Onem, G., et al.
2006 "Protection of Museum Items against Earthquake Shaking by Low-Cost Base-Isolation Devices," paper presented at the Fourth World Conference on Structural Control and Monitoring, La Jolla, Calif.

Podany, J. C.
1991 "Safeguarding a Collection and Building from Natural Hazards," *Perspectives on Natural Disaster Mitigation: Papers Presented at 1991 AIC Workshop* (Washington, DC), pp. 51–67.

Sungay, B., and N. Erturk
2004 "Seismic Conservation of Historical and Cultural Treasures of a World City: Sizing the Need and Formulating an Action Plan for the Museums of Istanbul, Turkey," *Final Report to the World Bank ProVention Consortium* (Istanbul).

Seismic Mitigation at the Asian Art Museum in San Francisco

Donna Strahan

Setting Seismic Mitigation as a Priority

The founding of the Asian Art Museum in 1964 was predicated on an extraordinary gift of seven thousand works of art donated to the people of San Francisco by Chicago industrialist Avery Brundage. The museum maintains one of the most comprehensive collections of Asian art in the world. Spanning six thousand years, the collection has a scope and breadth that enables the museum to provide an intensive historical education in all major cultural and religious traditions of Asia. Some of the highlights of the collection include: a comprehensive collection of Chinese jades, bronze objects, and rare painted Japanese scrolls and screens; ceramics from the third millennium B.C. to the present; major Buddhist sculpture; and superb decorative arts and textiles. In addition, it has the finest and most wide-ranging collection of Korean art outside of Korea; Himalayan paintings and major groups of sculpture; a comprehensive collection of Indian sculpture; Thai paintings and ceramics; Cambodian stone and bronze materials; and textiles, sculpture, jewelry, ceramics, terracotta works, and paintings from Indonesia, Burma (now Myanmar), Vietnam, and the Philippines. Over the years, the collection has grown to over sixteen thousand objects. It is now housed in a historic building, located in the Civic Center, which has recently been seismically retrofitted (fig. 1).

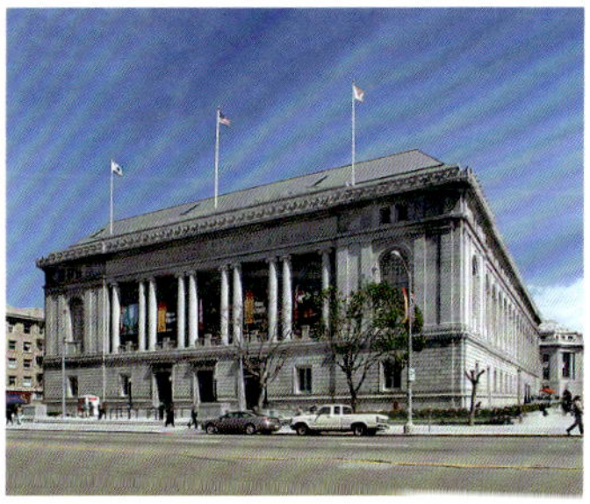

Figure 1
The Asian Art Museum's retrofitted building in San Francisco's Civic Center, just after opening in March 2003. Photo: Kaz Tsuruda, Asian Art Museum.

In its original building—opened in 1966 as a new wing of the de Young Museum in Golden Gate Park—the Asian Art Museum had difficulty establishing itself as an independent identity (fig. 2). Furthermore, the museum's collection had grown—it had nearly doubled in size by 1988—but the building that housed the museum was not up to the highest standards. Thus the museum began to search for a building of its own. Then in 1989, the old museum was rendered unstable by the Loma Prieta Earthquake, which measured 7.0 on the Richter scale (fig. 3). The need for a new home became pressing.

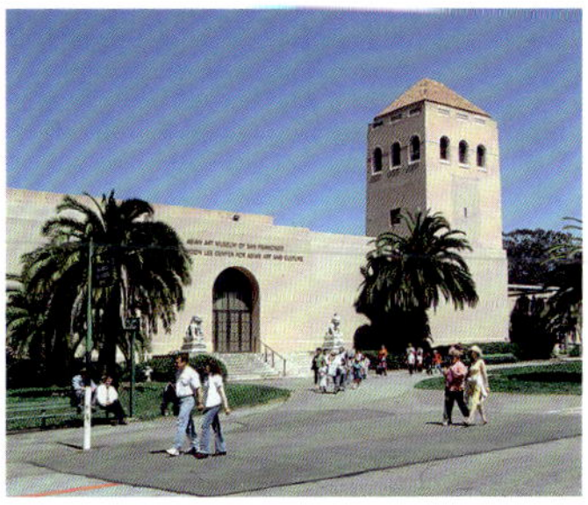

Figure 2
The Asian Art Museum in Golden Gate Park, just prior to the museum's move to the Civic Center and the demolition of this building. Photo: Kaz Tsuruda, Asian Art Museum.

The Loma Prieta Earthquake damaged the building and over forty objects in the museum's collection. It also damaged the older part of the building housing the de Young collection. As a necessary response to the earthquake damage, the museum hired Vince Avalos as a mount maker to create earthquake-proof exhibition mounts. Prior to this time, there had not been much attention to mounts or storage methods with respect to their role in damage mitigation, but the damage from the earthquake was so extensive that the museum realized that preventive measures had to be taken. The board and staff

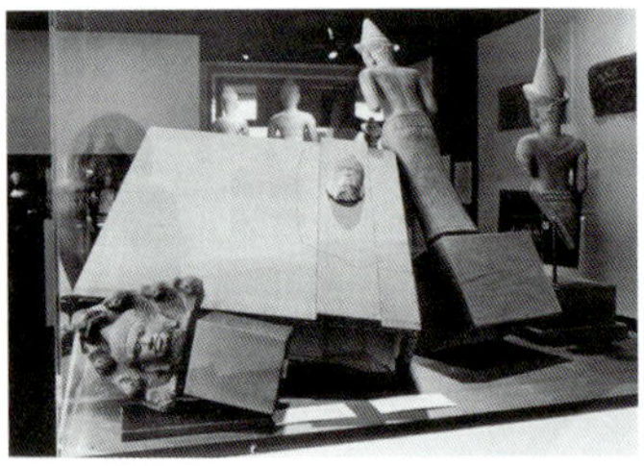

Figure 3
The 1989 Loma Prieta Earthquake caused extensive damage to sculpture and exhibition cases in the Golden Gate Park building. Photo: Asian Art Museum.

determined that a new seismically safe building was a critical need, and the city of San Francisco responded.

Seismic Retrofitting of the Building

In 1998 the city gave the Asian Art Museum the 1917 Beaux Arts building on the municipal plaza in downtown San Francisco that was formerly the city library. The old library building was originally constructed on concrete footings bearing on dense sand dunes (much of San Francisco is built on sand dunes). It was supported by a riveted steel frame encased in brick masonry infill and unreinforced concrete fireproofing. The infill acted as shear walls, providing limited resistance to lateral movement. The independent steel frame of the north wing had been simply bolted to the envelope of the building, and it had sustained the worst damage from the Loma Prieta Earthquake in 1989. In 1992 a seismic upgrading study recommended that substantial earthquake retrofitting of the building take place before any pubic use, and seismic isolation at the foundation, along with shear wall strengthening, was recommended.

The transformation of the old library building into the Asian Art Museum was envisioned by architect Gae Aulenti (working with the local design team of Hellmuth, Obata and Kassabaum; LDA Architects; and Robert B. Wong) and placed under the construction management team of Lem Construction (LCI) and DPR Construction. More than forty subcontractors were employed over the time that it took to complete the various stages of the project: protection of the historic finishes of the building, soft and hard demolition, and hazardous materials abatement, along with seismic retrofitting (including base isolation) of the historic building. Work began in earnest in 1999 (fig. 4).

Added to the challenge of retrofitting the building was the discovery of burial grounds from the Gold Rush period. The site of the library building had originally been the location of San Francisco's first city hall and city ceme-

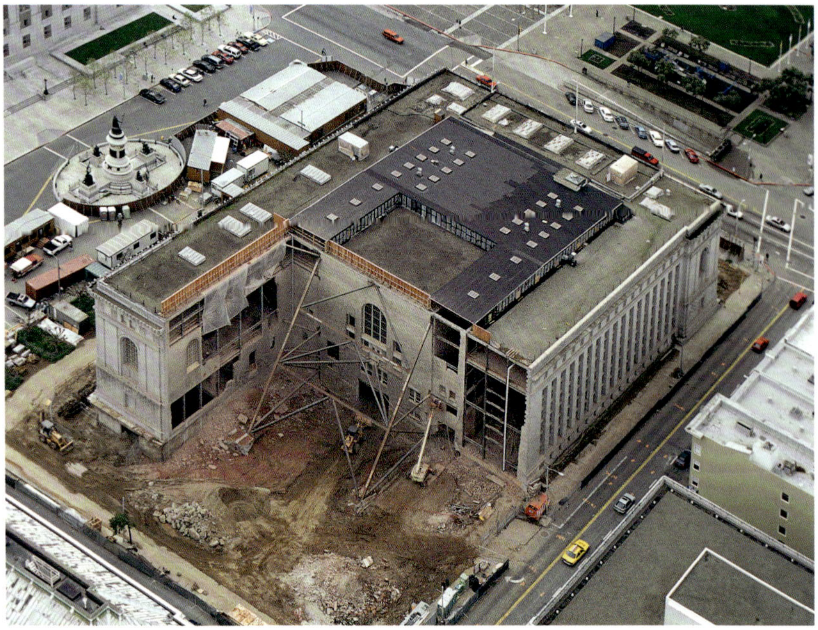

Figure 4
With temporary bracing in place, old, seismically unsound brick walls are removed from the Asian Art Museum's Civic Center building. Photo: Joseph Park, Perretti & Park Pictures.

tery. When the library was being built in 1917, all of the burials were supposed to have been transferred to a cemetery south of the city. However, during the renovation, more burials were discovered. Archaeologists were called in, and the remains were properly reburied elsewhere.

The firm Forell/Elsesser Engineers of San Francisco was charged with the building's seismic retrofitting. Paul Rodler, engineer of record, was given the museum's criteria for the structural design of the building. These criteria were: for a Richter scale magnitude 7 earthquake, there would be no loss of the collection in storage or on display; for a great earthquake (such as the earthquake of 1906), there would be no loss in storage and less than 1 percent loss of displayed materials. These criteria were quite stringent, as they far exceeded any other building codes at the time, including those of hospitals. This process, overseen by Dynamic Isolation Systems of Sparks, Nevada, began with the installation of temporary steel lateral bracing to provide a solid frame for the structure during this fragile and vulnerable stage of construction.

Seismic isolation involved the placement of base isolators, each weighing more than 2268 kg (5000 lb.), under load bearing walls and all pillars. Each base isolator is composed of forty-eight alternating layers of steel and rubber disks sandwiched between top and bottom steel plates (fig. 5). They were designed to absorb the quake energy resulting from earthquakes up to a Richter scale value of 8.0 and to equalize the resulting effects, so that the building would respond to these forces as a single unit. The base isolators are primarily designed for the lateral movement of an earthquake. They are very stiff for vertical earthquake movement, but the vertical movement is usually about one-third the intensity of the lateral movement.

The lower level was first excavated, with the area below the pillars prepared for the installation of base isolators (fig. 6). A new, structurally sound floor had to be constructed on which to set the base isolators. Once the floor

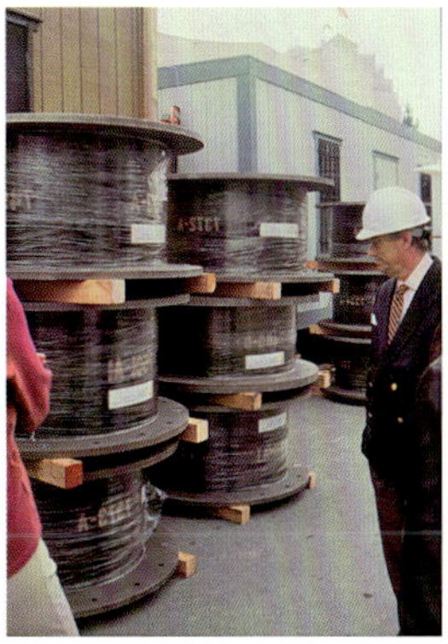

Figure 5
Stacks of base isolators before they were installed under the building. Photo: Asian Art Museum.

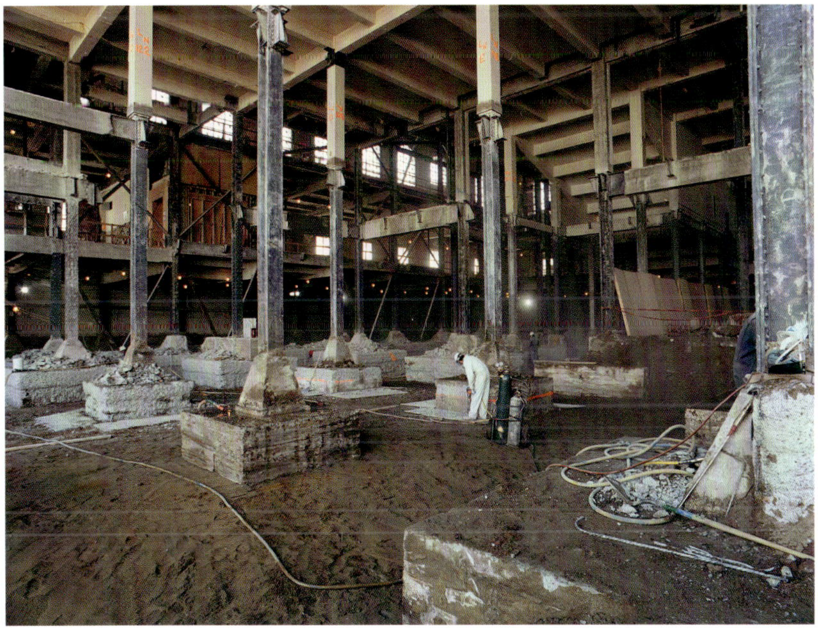

Figure 6
Pillar bases being prepared for the installation of the base isolators. Photo: Joseph Park, Perretti & Park Pictures.

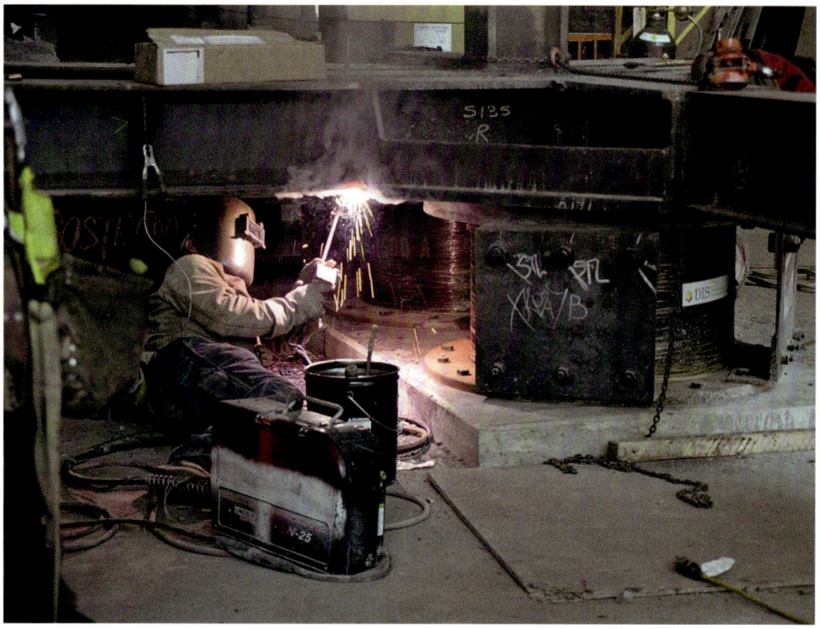

Figure 7
A partially completed base isolator during installation. The steel deck, framing, and connection to the existing building pillar are shown. Photo: Joseph Park, Perretti & Park Pictures.

was completed, a total of 204 base isolators were placed under all the pillars and walls of the building. In addition, all of the building's shear walls were heavily reinforced with rebar.

In order to install the isolators and give them room for movement in the event of a stress to the building, a moat—305 m (1000 ft.) long, 3 m (10 ft.) deep, and 1.5 m (5 ft.) wide—was dug around the perimeter of the building. Once the base isolators were installed, the moat was covered with panels that are hinged so that they pop up during an earthquake and allow the building to move.

The building load was transferred to the isolators after they were in place by releasing six bolts each from hundreds of individual locking plates (fig. 7). The base isolation and seismic retrofitting of the building were the most critical parts of the process, and they were the most time-consuming and expensive elements of the renovation. The process took three years and cost more than $50 million.

Earthquake Protection in Storage

The move also gave the museum an opportunity to both renovate and change its storage methods. In the new building, storage of art objects was once again located in the basement. This level of the building lost 1.2 m (4 ft.) of height because of the space taken up by the base isolators. In addition, the presence of the 179 pillars of the original building proved challenging when the new storage spaces were designed. Furthermore, there are access hatches for maintenance or replacement of the base isolators which could not be covered over with cabinets.

The museum received a National Endowment for the Humanities grant to purchase new storage cabinets. Delta Designs and Spacesaver Corporation designed a combination of fixed and compact cabinets that met the particular needs of the collection, including special cabinets for Asian scrolls and screens and Tibetan thangkas. During installation, all of the fixed cabinets were anchored and braced to the wall and floor with bolts and gussets. The compact shelving units were secured to their support rails to keep them from falling over in an earthquake.

This opportunity provided the time to rehouse much of the museum's collection. Scrolls were placed in individually fitted archival boxes, and painted screens were tied to individual panels mounted on wheels and then rolled into cabinets for easy access. Large flat paintings, such as thangkas and Thai paintings, were placed in flat drawers in closed cabinets. Three-dimensional objects with high centers of gravity were placed on their sides on Ethafoam lined shelves. Each object was surrounded by sandbags and given extra space to allow for vibrations (fig. 8). Custom-fit boxes in the drawers were made with Coroplast dividers padded with Ethafoam and acid-free tissue. Since the curators wanted easy access to large stone sculpture, any piece heavier than one person could handle was placed upright and tied down to its own mount (fig. 9).

Special Mounts for the Exhibitions

All of the new cases were constructed in the exhibition spaces. The wall cases were bolted to the building, and the freestanding cases were weighted

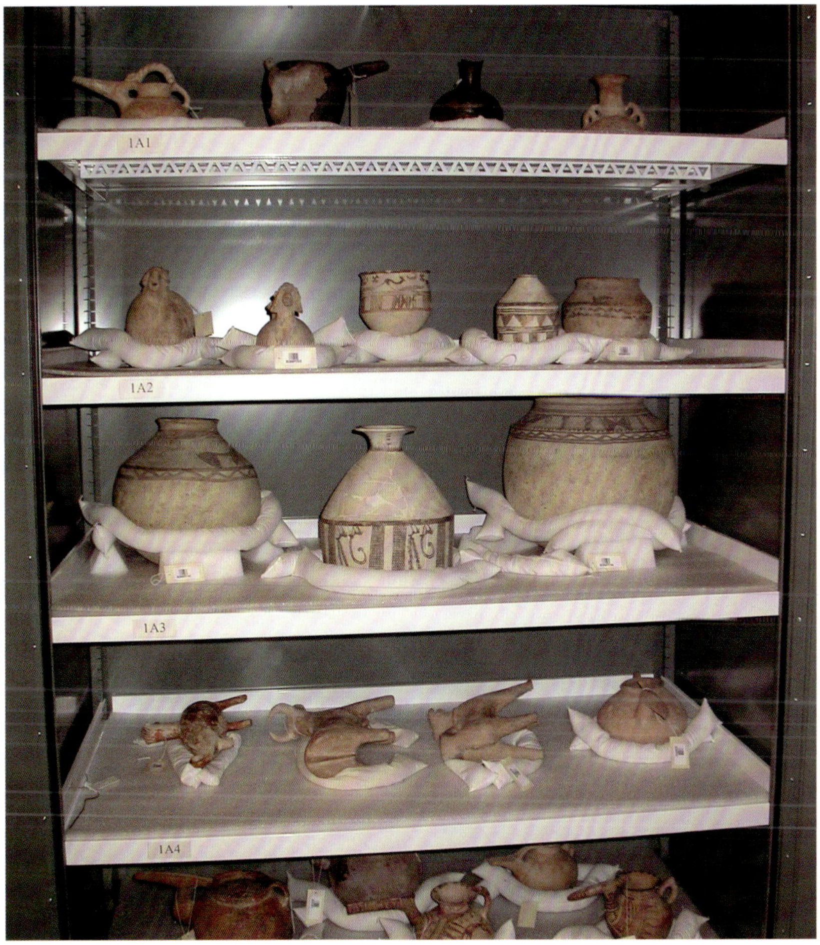

Figure 8
Ceramic storage cases in which sandbags protect objects from moving during earthquakes. Photo: Donna Strahan, Asian Art Museum.

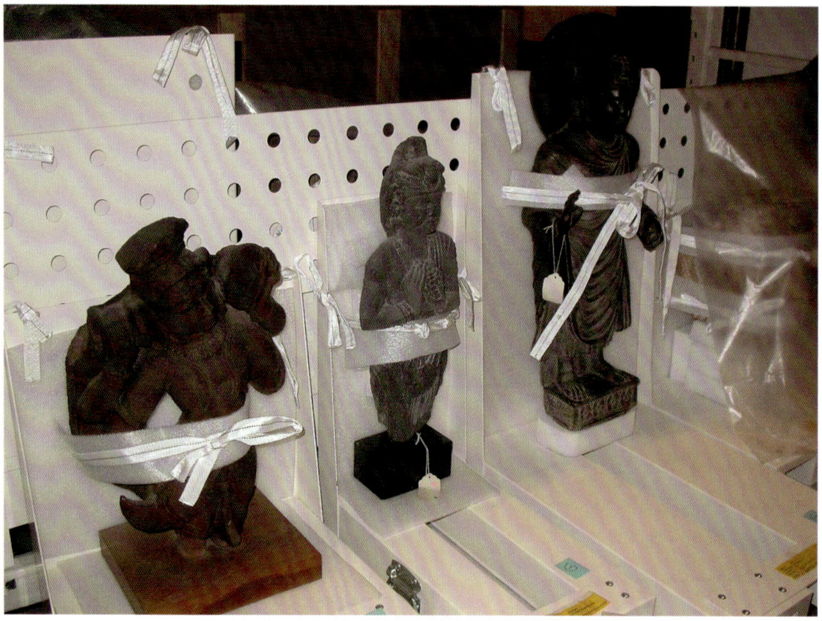

Figure 9
Large stone sculptures mounted in storage for easy viewing. Photo: Donna Strahan, Asian Art Museum.

with lead bricks to lower their centers of gravity and prevent them from toppling over during seismic activity.

Even though the building was base isolated, it was felt that earthquake mounts for individual objects were still necessary. Shake table tests had been done on individual object mounts, and it was determined that they were deemed reliable up to 0.4–0.6 g. If the building alone were made stronger then the objects inside, it would move more violently. Putting the building on base isolators allowed the structure the flexibility to move 0.76 m (2 ft., 6 in.) in any direction. Therefore, the acceleration of an earthquake moving into the building would be reduced by the isolators by as much as 4 to 5 times. It was felt that the object mounts could safely sustain such movement.

In order to accomplish the task of fabricating over a thousand new mounts, four mount makers worked full-time for a year and a half to prepare mounts for the objects going into the new galleries (fig. 10). Each mount had to be individually fitted to the object it secured. Mounts were fabricated for a wide range of objects, from small, lightweight items (fig. 11) to very heavy monumental objects. Top-heavy objects were weighted down with bags of sand to lower their centers of gravity. Monofilament was often used to tie objects to their mounts. Smaller objects were held down with microcrystalline wax (with the exception of lacquer objects, because of a concern about the potential interaction between the lacquer surfaces and the wax). Any objects on loan to the museum, as well as all the traveling exhibitions accepted by the museum, receive the same type of custom mounts when they are displayed at the Asian Art Museum.

Conclusion

The process of seismic retrofitting has brought the building up to exacting standards, so it is expected to sustain almost no damage in the event of an earthquake. It has been three years since the base isolators were installed, and thus

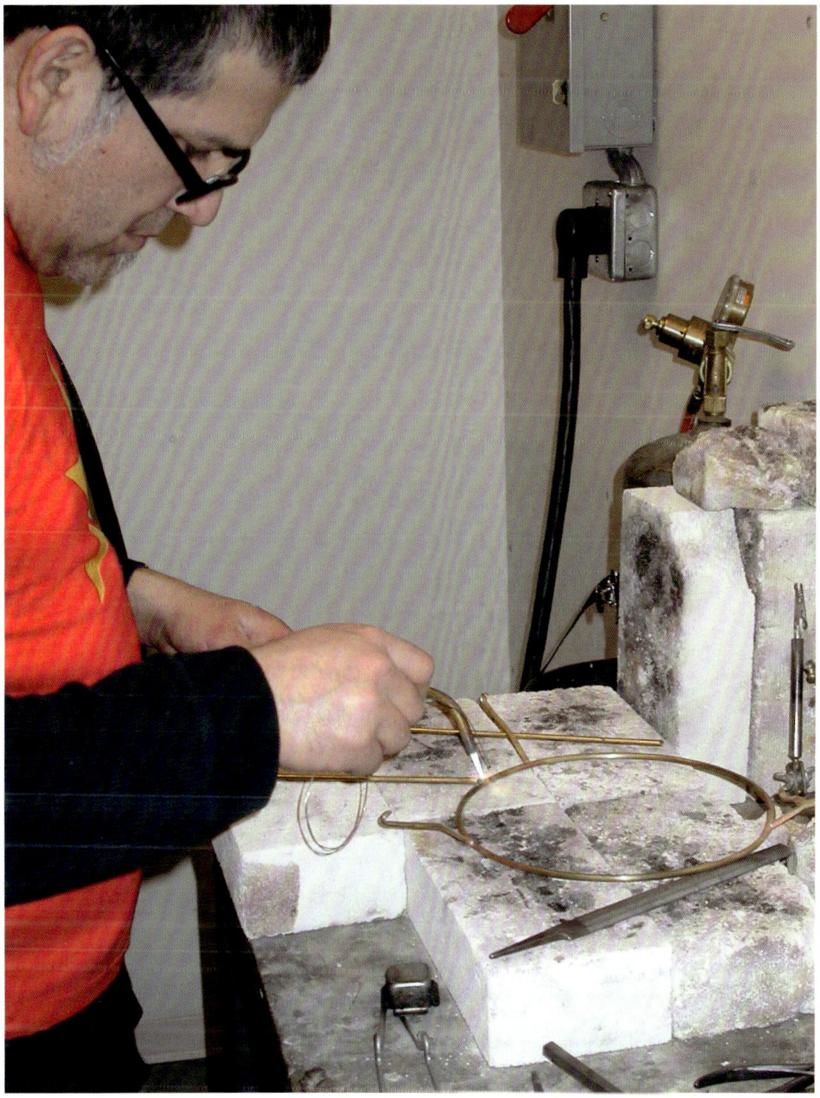

Figure 10
Mount maker Vince Avalos creating a seismic mount for an object. Photo: Donna Strahan, Asian Art Museum.

far, there have been a few tremors but no major earthquakes. During this time, there has been no need for any major maintenance of the base isolator system.

The new building opened in March 2003. It provides greatly improved facilities for museum visitors and staff and has attractive new spaces for the collection and for exhibitions. The new Asian's 2964 m^2 (29,000 sq. ft.) of space in thirty-three permanent collection galleries enable approximately 20 percent of the collection to be on view at any one time—double the space in the old building. The building's 790 m^2 (8500 sq. ft.) of temporary exhibition space will host a program of changing exhibitions. The museum now houses its important collection of sixteen thousand works of art in a secure, safe, and stable environment. This seismic retrofitting project was an enormous undertaking —one that clearly demonstrated that planning, conservation, presentation, and preservation are crucial elements of a coherent, comprehensive, and successful process of damage mitigation and preventive conservation.

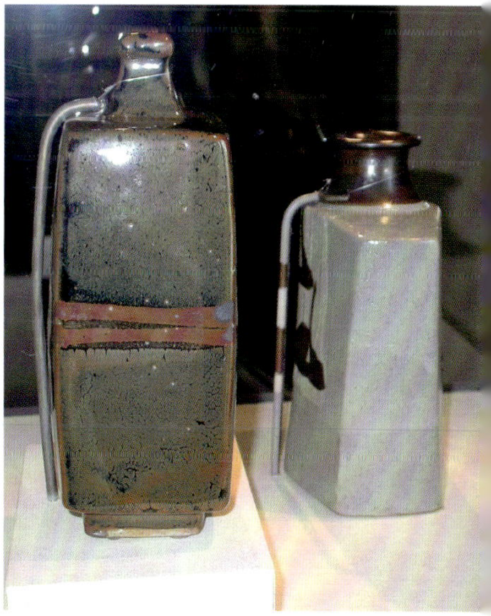

Figure 11
Japanese ceramics held with earthquake mounts made of metal and monofilament. Photo: Donna Strahan, Asian Art Museum.

Case Study: A Review of Seismic Mitigation Efforts for Museums in Greece

Constantine Spyrakos

Ioannis Koutromanos

Charilaos Maniatakis

Introduction

The principal objective of this paper is to provide a brief description of past and present efforts to protect art objects in Greek museums from earthquake-induced damage. Greece, a country celebrated for its cultural heritage, is also a country with one of the highest seismicities in the world. Every two years there is an earthquake with a magnitude of at least 6 on the Richter scale. Fortunately, most of the earthquakes occur at the bottom of the sea. However, there have been a number of strong earthquakes that have occurred near or on the mainland which have caused loss of life and severe damage to engineered structures. Representative strong earthquakes that have occurred in Greece during the last eighty years are listed in table 1, in terms of the surface wave magnitude scale M_S, as defined by Gutenberg (Gutenberg 1945). Table 1 indicates that strong earthquakes have occurred in almost every part of Greece. Damage to both museums and artifacts have been observed in many cases at these seismically struck areas. Figure 1 presents an example of dramatic art object failures caused by overturning and impact during the 1981 Alkyonides (Corinth) Earthquake at the Perachora Museum (Papazachos and Papazachou 1997), which was located within a few miles from the epicenter.

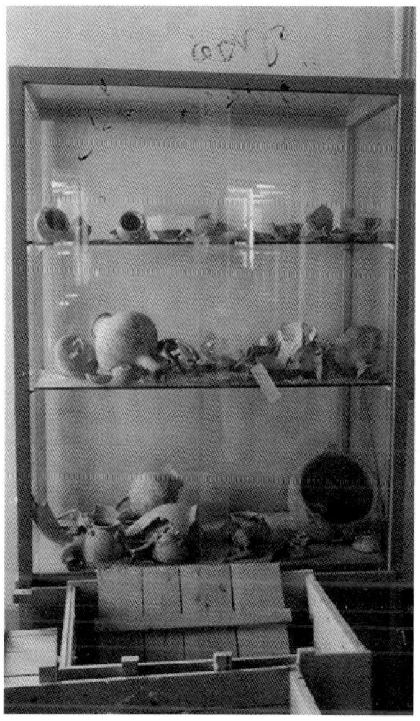

Figure 1
Failures of artifacts at the Perachora Museum, caused by overturning and impact, during the 1981 Alkyonides (Corinth) Earthquake.

The Athens Earthquake of 1999 not only caused severe damage to numerous engineered structures but also damaged artifacts and statues in museums. The Athens Earthquake, the magnitude of which was defined as $M_L = 5.4$ in the local magnitude scale (Richter 1935) and as $M_S = 5.9$ in the surface wave magnitude scale (Gutenberg 1945), was originated by a normal fault rupture at a focal depth of almost 12 km (7.4 mi.) and at a distance of 18–20 km (11.2–12.4 mi.) from the historical center of Athens. The seismic moment, M_o (Aki 1966), which is a quantity proportional to the energy released during an earthquake measured in units of newton meters (Nm), was estimated as $7.8 \cdot 10^{17}$ Nm by the United States Geological Survey and as $1.2 \cdot 10^{18}$ Nm by the Harvard Centroid Moment Tensor (CMT) database. The event was unexpected, since it occurred in a region traditionally considered of low seismicity, and there was no mapped fault identified as active in the epicentral area. The earthquake mainly hit the northwestern Athens suburbs, which are close to the causative fault. More than eighty buildings collapsed, including industrial facilities, causing 143 deaths, while the injured numbered as many as two thousand. About sixty thousand buildings suffered different levels of damage, from slight to heavy. The tangible losses were roughly estimated at about US$3 billion, and from this point of view, it was the worst natural disaster in modern Greek history.

Table 1
Major earthquakes in Greece (M_S refers to the surface wave magnitude scale as defined by Gutenberg [1945]; the initials in the last column—N, C, and S—indicate northern, central, and southern parts of Greece; the Aegean and Ionian seas are indicated with A and I, respectively).

	Epicenter	Date	M_S	Location
1	Ierissos	September 30, 1932	7.0	N
2	Thessaloníki	June 20, 1978	6.5	N
3	Athens	September 7, 1999	5.9	C
4	Alkyonides	February 24, 1981	6.7	C
5	Aigion	June 15, 1995	6.0	S
6	Kalamata	September 13, 1986	6.0	S
7	Sparti	August 30, 1926	7.2	S
8	Karpathos	February 9, 1948	7.1	A
9	Argostoli	August 12, 1953	7.2	I

The works of art in the Perachora Museum were severely damaged by overturning or sliding and impact. The main reason for their failure was the fact that they were unanchored.

Mitigation Measures

Code Requirements One of the primary reasons for failures of building and artifacts is the fact that no seismic code was enforced in Greece until 1959. However, seismic codes were enforced locally at certain regions in Greece at the beginning of the nineteenth century. The current Greek seismic code (Earthquake Planning and Protection Organization 2000) requires use of elastic spectra for artifacts with ordinates proportional to (1) the maximum ground acceleration and (2) the importance factor $\gamma_I = 1.30$, which amplifies the design value by 30%. The code also recognizes the fact that the acceleration increases with height. It allows for either a dynamic analysis or an equivalent static analysis. For the latter, the seismic load is applied as a static load at the center of gravity of the object, amplified by a factor that is proportional to the ground acceleration, depends on the height of the object from the foundation, and depends on the fundamental period of both the object and the structure. As shown in figure 2, the design horizontal force on the object is given by

$$H_p = \frac{\frac{A}{g} \cdot \beta \cdot \left(1 + \frac{z}{H}\right) \cdot W_p \cdot \gamma_p}{q_p} \tag{1}$$

Where H_p is the horizontal design seismic load, A is the amplitude of the maximum ground acceleration, g is the acceleration of gravity, z is the height of the object support level, H is the total building height, W_p is the object weight, γ_p is the importance factor of the object, and q_p is a "behavior" factor that is taken equal to 1 for the linear (undamaged) behavior of the artifact.

The interaction coefficient β is given by

$$\beta = \frac{2}{1 + \left(1 - \frac{T_p}{T_{str}}\right)^2} \tag{2}$$

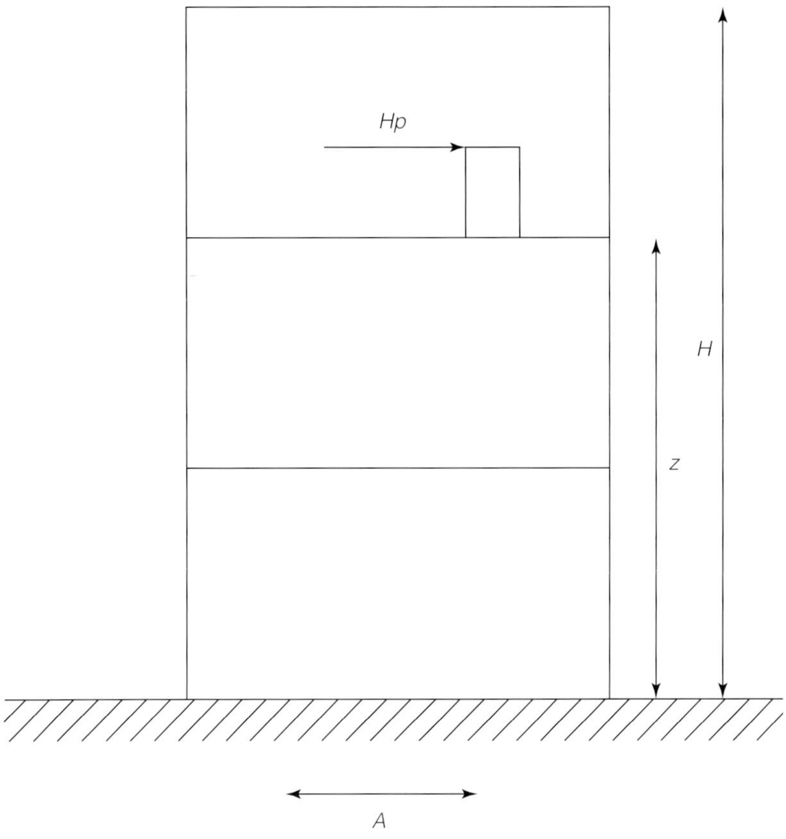

Figure 2
Horizontal seismic force on an artifact placed at a height z from the foundation of a frame structure (H_p = design horizontal seismic load, H = total building height, and A = amplitude of the maximum ground acceleration).

where T_p is the object fundamental period on a fixed base, and T_{str} is the structure's fundamental eigenperiod. It can easily be deduced that the maximum value of β is equal to 2. This amplification occurs for values of T_p close to or equal to T_{str} and could be insignificant for artifacts placed in rigid structures.

In an effort to provide museum personnel with easily applied directives, in 2002 the Greek Ministry of Culture published a series of guidelines concerning the design of support systems and display cases, as well as appropriate post-earthquake measures for artifacts damaged by earthquakes (Ghiossi et al. 2002).

Experimental Investigation A recent experimental investigation is presented in a report by Zampas and colleagues (2004) for the design of support systems in the Benaki Museum in Athens. Full-scale shake table tests were performed at the National Technical University of Athens Laboratory of Earthquake Engineering, in order to investigate the response of two replica artifacts and a replica vase, supported by a slender base. The artifacts were considered representative for Greek museums today. The study also aimed to evaluate the performance of a pin support system. The system was subjected to different seismic excitations, both artificial and representative of recorded Greek earthquakes.

The tests revealed a magnification of the earthquake accelerations as one moves from the base to the top of an object. Furthermore, the support pins exhibited permanent flexural deformations, which in turn resulted in tilting of the artifacts.

In general, it can be stated that experimental investigation into the seismic response of artifacts in Greece is limited. Only recently, with art object failures during the 1999 Athens Earthquake, has attention been drawn to the area of experimental testing.

Base Isolation The application of new mitigation technologies could effectively protect against earthquake-induced failures. Such a technique is base isolation. By means of special devices termed isolators, the response of the art object is isolated from the ground seismic motion. Isolators are made of material like rubber, and they are characterized by low horizontal stiffness. During an earthquake, the seismic displacement is concentrated in the isolator, while the art object moves as a rigid body without the development of significant deformations and stresses. Base isolation greatly reduces the transfer of seismic acceleration forces to the art objects and thus results in a lower potential for damage. Despite the fact that base isolation is a method that can be considered cost-effective in cases of important statues or artifacts, its application in Greece is limited to a few cases, the most important being that of the Hermes statue in the Archaeological Museum of Olympia (see the paper by Koumousis, "Seismic Isolation System for the Statue of Hermes at the Archaeological Museum of Olympia," in the present volume).

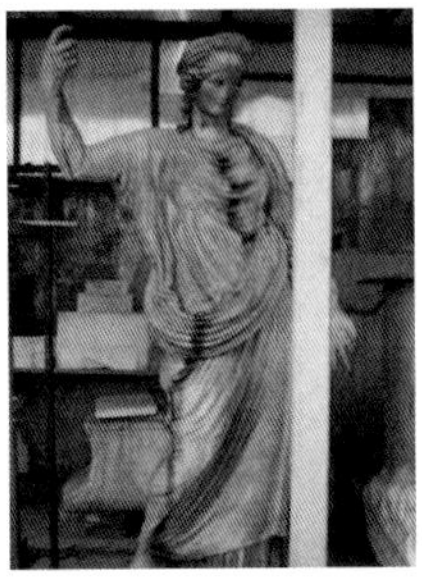

Figure 3
Replica statue that was modeled with finite element analysis (FEA).

Analytical Studies

The authors' ongoing research for the National Archaeological Museum of Athens, a description of which follows, is an example of the analytical investigation of the seismic response of artifacts. The research involves the analysis and design of six replica statues made either of clay or of a composite material. The task also involves the design of a metal frame to hold the hollow replicas. The replicas will replace six statues that were partially damaged during the 1999 Athens Earthquake and which have been removed from the roof of the main entrance to the National Archaeological Museum of Athens. The research project also involves the seismic hazard assessment of representative artifacts. The present study gives the results for ten artifacts—five located on the ground floor and five on the first floor of the two-story building of the National Archaeological Museum.

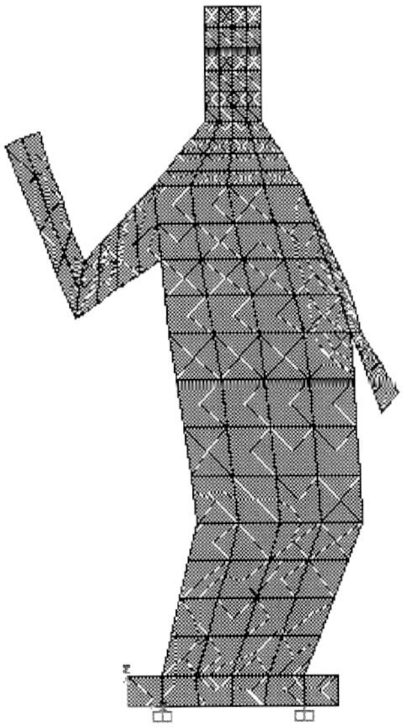

Figure 4
Finite element model of a clay replica statue at the National Archaeological Museum of Athens.

Analysis of Statues The finite element analysis (FEA) was employed to model the six replica statues (Spyrakos 1994; Spyrakos and Raftoyannis 1997). The representative statue of Athena is shown in figure 3. The finite element model of the statue is depicted in figure 4. The method is applied to study the dynamic response of the replica during an earthquake, while at the same time contributing to the identification of failure modes, which cannot be captured by a simple rigid-body dynamics model. The analysis, which included seismic loads specified by the Greek seismic code corresponding to a design horizontal ground acceleration of 0.24 g, revealed areas of the statue where high stress concentrations could develop and cause damage either to the structure or to its mounts. Weak areas or areas of stress concentrations appeared in regions with abrupt changes in geometry, such as the connections of the arms with the statue body and the statue ankles (see fig. 4). Such regions susceptible to high stresses could be strengthened either by means of larger local thickness or by means of appropriate strengthening metal reinforcement.

Analysis of Rigid Artifacts The analysis of the artifacts involved the estimation of the critical peak ground acceleration (PGA) and peak ground velocity (PGV) that could cause rocking and overturning, respectively, for the objects shown in figure 5. Sliding was not examined, since sliding requires experimental measurement of the friction coefficient at the interface of the artifact and its supports. The analysis was based on the methodology introduced by Agbabian, Masri, and Nigbor (1990) for the evaluation of seismic mitigation measures at the J. Paul Getty Museum, as well as an analytical study of Spyrakos and Nikolettos (2005). Assuming that the earthquake response of an artifact can be adequately described using rigid body kinematics, three failure modes are identified: rocking, sliding, and overturning.

The first two failure modes are controlled by the PGA developed during the earthquake, while the third one is controlled by the PGV. According to Spyrakos and Nikolettos (2005), the overturning stability of a flexible structure is determined by a criterion relating the structure's flexibility, inertia, and maximum velocity:

$$\left[T_K + K_e\left(\frac{m\varphi_1}{m^*\omega}\right)^2\right]S_{vo}^2 - W\frac{m\varphi_1^2\alpha}{m^*H\omega}S_{vo} - \frac{1}{2}WR\alpha^2 = 0 \quad (3)$$

where S_{vo} is the spectral velocity required to cause overturning, W is the total weight of the structure, B is the object width, H is the object height, g is the acceleration of gravity, $R = 0.5(B^2 + H^2)^{1/2}$, and $\alpha = B/H$. (Note that values in the tables and equations in this paper are given in metric units.) The other parameters are explained in detail in the reference paper by Spyrakos and Nikolettos (2005). For the limit case of a rigid structure, their criterion reduces to

$$\frac{B}{H} = \frac{S_{vo}}{\sqrt{gR}} \quad (4)$$

The criteria for the three modes of failure (Agbabian, Masri, and Nigbor 1990) are:

$$\begin{aligned} &A/g > B/H' && \text{(rocking)} \\ &V > 10B/\sqrt{H'} && \text{(overturning, } V \text{ in cm/sec., } B \text{ and } H \text{ in cm)} \\ &A/g > C && \text{(sliding)} \end{aligned} \quad (5)$$

where A and V are the maximum horizontal acceleration and velocity at the base of the object, respectively, B is the object base width, C is the friction coefficient, and H' is an "equivalent object height" defined for complex object geometries. This "equivalent height" depends on the art object area, second moment of inertia about a horizontal axis at the ground level, and vertical distance of center of gravity from the ground level. The equations (5) define the thresholds of horizontal acceleration (A) and velocity (V) that can cause failure.

The PGA and PGV values required for rocking or overturning are listed in table 2. An amplification factor of 2.5 is used for the artifacts located on the first floor—that is, the critical PGA and PGV values from the failure criteria are divided by a factor equal to 2.5.

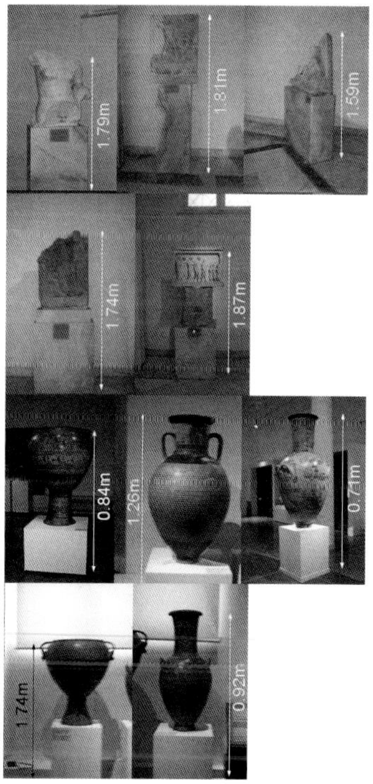

Figure 5
Artifacts at the National Archaeological Museum of Athens for which a seismic hazard assessment was carried out. The top two rows show objects located on the ground floor; the bottom two rows show objects located on the first floor.

Table 2
Critical values of *PGA* and *PGV* for rocking and overturning, respectively, of the National Archaeological Museum of Athens artifacts.

Artifact	*PGA* (g)	*PGV* (cm/sec.)
1	0.189	36.11
2	0.139	27.18
3	0.286	51.33
4	0.168	30.53
5	0.122	23.17
6	0.117	13.12
7	0.231	31.56
8	0.265	27.04
9	0.128	18.79
10	0.135	14.96

Hazard Estimation

A procedure to estimate the seismic risk involves the correlation of the *PGA* and *PGV* with the epicentral distance and the earthquake intensity, usually expressed by the modified Mercalli intensity (*MMI*) scale with appropriate attenuation relations. Knowledge of *PGA* and *PGV*, in conjunction with the knowledge of active faults and historical seismicity in the area of interest, allows for a preliminary assessment of the need to resort to further mitigation measures.

In the present study, the empirical attenuation relations by Theodulidis and Papazachos (1992) were employed. These expressions relate *PGA* and *PGV* with the epicentral distance and surface wave magnitude M_S, according to the following equations:

$$\ln(PGA) = 3.88 + 1.12 \cdot M_S - 1.65 \cdot \ln(R + 15) + 0.41 \cdot S \quad (6a)$$

$$\ln(PGV) = -0.79 + 1.41 \cdot M_S - 1.62 \cdot \ln(R + 10) - 0.22 \cdot S \quad (6b)$$

where *R* is the epicentral distance in km, and *S* is a Boolean parameter that is equal to 1 for rock sites and equal to 0 for alluvium sites. The *PGA* and *PGV* are expressed in cm/sec.^{2} and cm/sec., respectively.

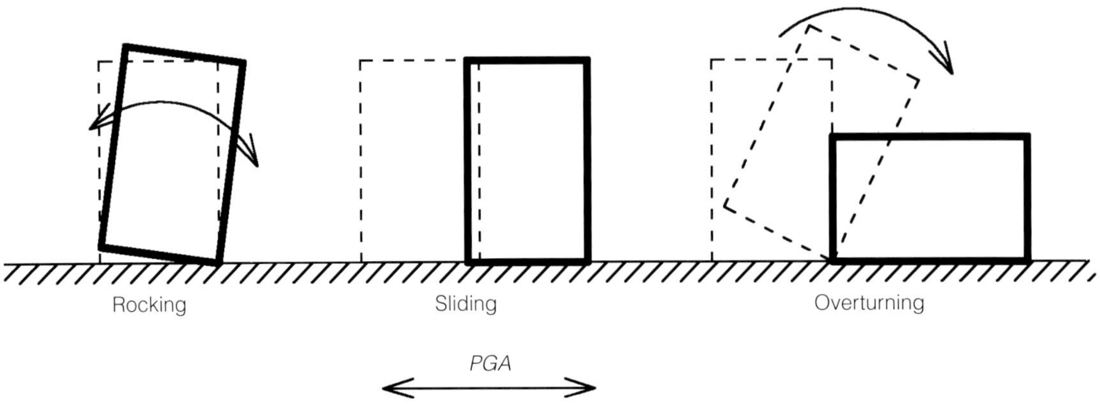

Figure 6
Failure modes for rigid bodies subjected to base excitations (*PGA* = peak ground acceleration).

The respective relations for the *MMI* are:

$$\ln(PGA) = 0.28 + 0.67 \cdot MMI + 0.42 \cdot S \quad (7a)$$

$$\ln(PGV) = -3.02 + 0.79 \cdot MMI - 0.04 \cdot S \quad (7b)$$

In order to demonstrate the application of the methodology, the epicentral distances were estimated for two different earthquake magnitudes, $M_S = 5.5$ and $M_S = 6.5$, and for two soil conditions—that is, rock and alluvium deposits as listed in tables 3 and 4, respectively.

Table 3 can be explained in terms of the results for the first artifact (no. 1). Based on the criteria established by Agbabian, Masri, and Nigbor (1990), the epicentral distance at which a peak ground acceleration of at least $PGA = 0.19$ g, which can cause rocking of the artifact, is calculated and shown. Using equation 6a for a rock site ($S = 1$) and a surface wave magnitude $M_S = 5.5$, the maximum epicentral distance at which artifact no. 1 could be placed in

Table 3
Estimated values of epicentral distance for rock sites for either rocking (r) or overturning (o) of the National Archaeological Museum of Athens artifacts.

	Epicentral distance (km)	
Artifact	$M_S = 5.5$	$M_S = 6.5$
1	8.76 (r)	31.84 (r)
2	13.57 (r)	41.33 (r)
3	3.47 (r)	21.41 (r)
4	10.53 (r)	35.34 (r)
5	15.98 (r)	46.08 (r)
6	16.76 (r)	47.61 (r)
7	6.05 (r)	26.50 (r)
8	4.34 (r)	23.14 (r)
9	15.07 (r)	44.29 (r)
10	14.09 (r)	42.35 (r)

Table 4
Estimated values of epicentral distance for alluvium sites for either rocking (r) or overturning (o) of the National Archaeological Museum of Athens artifacts.

	Epicentral distance (km)	
Artifact	$M_S = 5.5$	$M_S = 6.5$
1	3.53 (r)	21.54 (r)
2	7.29 (r)	28.94 (r)
3	—	13.40 (r)
4	4.91 (r)	24.26 (r)
5	9.17 (r)	32.64 (r)
6	9.77 (r)	33.84 (r)
7	1.42 (r)	17.37 (r)
8	—	14.75 (r)
9	8.46 (r)	31.24 (r)
10	7.69 (r)	29.73 (r)

order to be excited by a seismic motion with a ground acceleration of PGA = 0.19 g is R = 8.76 km. Using the same procedure, the epicentral distance shown in table 3 at which a peak ground velocity PGV = 31.84 cm/sec. can cause overturning of the artifact, according to the criteria proposed by Agbabian, Masri, and Nigbor (1990), is determined. For a rock site (S = 1) and a surface wave magnitude M_S = 5.5, equation 6b yields a negative maximum epicentral distance at which artifact no. 1 could be placed in order to be excited by a seismic motion with a velocity of at least PGV = 36.11 cm/sec. The negative value implies that it is impossible for the artifact to overturn, no matter how close to the epicenter it is placed, for a seismic motion with M_S = 5.5. Applying equation 6a for a seismic motion with a surface wave magnitude M_S = 6.5, the epicentral distance at which rocking is expected (PGA = 0.19 g) results in R = 31.84 km, while application of equation 6b results in R = 6.78 km for the epicentral distance at which overturning is expected (PGV = 36.11 cm/sec.). For each case, the epicentral distance was set equal to the maximum of the epicentral distances corresponding to either rocking or overturning. Applying the procedure for the ten artifacts, it is found that rocking (r) is the failure mode, as shown in table 3. Since, according to the attenuation relations, both PGA and PGV are reduced with increasing distances, the largest distance corresponds to the failure mode that occurs first. In table 4 the same procedure is applied for alluvium site conditions. Notice that also in this case, rocking is the failure mode for all artifacts. For rock soil conditions and an M_S = 5.5 earthquake, the estimated epicentral distance is rather small, ranging from 3.5 to 16 km for the ground floor, and from 4.3 to 16.8 km for the first floor. As expected, the epicentral distances increase substantially for M_S = 6.5, ranging from 21.4 to 46.1 km and from 23.1 to 47.6 km, respectively. For an M_S = 5.5 earthquake and an alluvium site, the attenuation relations failed to provide a positive distance for artifacts 3 and 8. For this reason no distance is assigned to these objects (see table 4). The epicentral distances for an M_S = 5.5 earthquake and alluvium site conditions range from 3.5 to 9.2 km for artifacts placed on the ground floor and from 1.4 to 9.8 km for artifacts placed on the first floor. The epicentral distances for an M_S = 6.5 earthquake and alluvium soil conditions range from 13.4 to 32.6 km for the ground floor and 14.7 to 33.8 km for the first floor

One may argue that the empirical relations employed in this study are not the most appropriate, since they have not been developed with data from near-source earthquakes, as elaborated by several researchers—e.g., Spyrakos, Maniatakis, and Taflambas 2005. For this reason, the epicentral distances required for overturning failure were also evaluated with the empirical relation developed by Rodriguez-Marek (Stewart et al. 2001).

$$\ln(PGV) = 2.44 + 0.50 \cdot M_S - 0.41 \cdot \ln(r^2 + 3.93^2) \qquad (8)$$

where r is the closest distance to the fault in km, and PGV is given in cm/sec. The fault distances, as calculated by equation 8, for M_S = 5.5 and M_S = 6.5 are presented in table 5. Since equation 7 includes PGV values, the epicentral distances shown in table 5 refer only to the overturning (o) failure type.

With equations 7a and 7b and through application of the procedure described for tables 3, 4, and 5, the minimum MMI is determined for the PGA and PGV values listed in table 2. The values shown in table 6 refer to the mini-

	Distance from rupture (km)	
Artifact	$M_S = 5.5$	$M_S = 6.5$
1	5.87 (o)	12.39 (o)
2	9.19 (o)	17.96 (o)
3	2.39 (o)	7.50 (o)
4	7.73 (o)	15.47 (o)
5	11.49 (o)	21.99 (o)
6	23.97 (o)	44.53 (o)
7	7.34 (o)	14.81 (o)
8	9.26 (o)	18.08 (o)
9	15.17 (o)	28.57 (o)
10	20.32 (o)	37.88 (o)

Table 5
Estimated values of rupture distance for overturning (o) of the National Archaeological Museum of Athens artifacts, referencing the Rodriguez-Marek (2000) relations (Stewart et al. 2001).

	MMI	
Artifact	Rock	Alluvium
1	VII (r)	VIII (r)
2	VII (r)	VII (r)
3	VIII (r)	VIII (r)
4	VII (r)	VIII (r)
5	VII (r)	VII (r)
6	VII (r)	VII (r)
7	VII (r)	VIII (r)
8	VIII (r)	VIII (r, o)
9	VII (r)	VII (r)
10	VII (r)	VII (r)

Table 6
Critical modified Mercalli intensity (*MMI*) for rock and alluvium sites.

mum *MMI* capable of causing either rocking or overturning of the artifacts. As expected, rocking is the first failure mode observed, with the exception of artifact no. 8, for which an earthquake intensity of *MMI* = VIII can cause either rocking (r) or overturning (o) when alluvium soil conditions are considered. It was found that an *MMI* of at least VII is required for both ground-floor and first-floor artifacts, as listed in table 6. Low *MMI* implies that a small-to-moderate earthquake may cause damage to the artifacts. However, rocking is not always equivalent to failure, since impact with neighboring objects is not always observed; moreover, the value of the amplification factor of 2.5 used in the calculations was rather conservative. Nevertheless, the results indicate the much higher vulnerability of artifacts placed at higher elevations in flexible buildings.

Conclusions

Seismic mitigation for art objects has been poorly addressed in several countries. Specific measures have been taken in Greece, especially after the recent 1999 earthquake. Innovative techniques such as base isolation have been applied in certain special instances. However, the experimental investigation of the behavior of art objects has been limited. It is imperative that a preliminary seismic risk analysis be carried out for every museum in Greece and that a strategic plan for earthquake mitigation measures be developed, in order to avoid failures of statues and artifacts in future earthquakes.

One factor that should also be taken into account is the potential occurrence of near-source earthquakes—that is, earthquakes with epicenters located near a site of interest, within about 10 to 30 km (about 6–19 mi.) (Spyrakos, Maniatakis, and Taflambas 2005; 2007). The strong motion characteristics of the near-source zone are greatly influenced by the fault rupture mechanism and the rupture propagation direction with respect to the site's location, as well as by permanent displacement phenomena attributed to the fault slip. When the fault rupture propagates towards the site of interest, a seismic energy concentration occurs, resulting in the occurrence of a velocity pulse in the direction normal to the fault plane (Somerville et al. 1997). Since the velocity pulse is expected to be characterized by medium- to large-period values, it results in increased seismically induced deformations to relatively flexible structures.

One should anticipate larger seismic deformations for art objects like tall slender statues or hanging items, or even for base isolated museums and statues. Thus, it is imperative that the effects of near-source strong motions are taken into account, through appropriate analytical and experimental investigation.

References

Agbabian, M. S., S. F. Masri, and R. L. Nigbor
1990 *Evaluation of Seismic Mitigation Measures for Art Objects* (Los Angeles).

Aki, K.
1966 "Generation and Propagation of G Waves from the Niigata Earthquake of June 16, 1964," *Bulletin of the Earthquake Research Institute* 44, pp. 23–88.

Earthquake Planning and Protection Organization [OASP]
2000 *Greek Seismic Code EAK2000* [in Greek] (Athens).

Ghiossi, S., M. Krini, T. Maragou, and P. Prokos
2002 *Earthquakes and Ancient Artifacts: Prevention and Post-Earthquake Mitigation Measures* [in Greek] (Athens).

Gutenberg, B.
1945 "Amplitudes of Surface Waves and Magnitudes of Shallow Earthquakes," *Bulletin of the Seismological Society of America* 35, pp. 3–12.

Papazachos, B., and C. Papazachou
1997 *The Earthquakes of Greece* (Thessaloníki).

Richter, C.
1935 "An Instrumental Earthquake Magnitude Scale," *Bulletin of the Seismological Society of America* 25, pp. 1–32.

Somerville, P. G., N. F. Smith, R. W. Graves, and N. A. Abrahamson
1997 "Modification of Empirical Strong Ground Motion Attenuation Relations to Include the Amplitude and Duration Effects of Rupture Directivity," *Seismological Research Letters* 68, no. 1 (Jan./Feb.).

Spyrakos, C. C.
1994 *Finite Element Modeling in Engineering Practice* (Pittsburgh).

Spyrakos, C. C., C. A. Maniatakis, and J. Taflambas
2005 "Critical Evaluation of Near-Field Seismic Records in Greece," *Earthquake Resistant Engineering Structures*, C. A. Brebbia, D. E. Beskos, G. D. Manolis, and C. C. Spyrakos, eds. (Southampton), pp. 53–62.

———
2007 "Evaluation of Near Source Seismic Records Based on Damage Potential Parameters," *Soil Dynamics and Earthquake Engineering* (in press).

Spyrakos, C. C., and G. S. Nikolettos
2005 "Overturning Stability Criteria for Flexible Structures to Earthquakes," *Journal of Engineering Mechanics* 131, no. 4 (April), pp. 349–58.

Spyrakos, C. C., and J. Raftoyiannis
1997 *Linear and Nonlinear Finite Element Analysis in Engineering Practice* (Pittsburgh).

Stewart, J. P., S. J. Chiou, J. D. Bray, R. W. Graves, and P. G. Somerville
2001 *Ground Motion Evaluation Procedures for Performance-Based Design*, PEER [Pacific Earthquake Engineering Research Center] Report 2001/09 (Berkeley).

Theodulidis, N. P., and B. C. Papazachos
1992 "Dependence of Strong Ground Motion on Magnitude-Distance, Site Geology and Macroseismic Intensity for Shallow Earthquakes in Greece: I, Peak Ground Acceleration, Velocity and Displacement," *Soil Dynamics and Earthquake Engineering* 11, pp. 387–402.

Zampas, K., K. Chatziantoniou, A. Galanou, and A. Dogani
2004 "Protection of Ancient Artifacts against Earthquakes" [in Greek], research project final report, Earthquake Planning and Protection Organization [OASP], Athens.

Seismic Isolation System for the Statue of Hermes at the Archaeological Museum of Olympia

Vlasis Koumousis

Introduction

A seismic isolation system for the statue and base of the Hermes of Praxiteles at the Archaeological Museum of Olympia was designed and installed. The Ministry of Culture of Greece funded this research project, which was supervised by the Directorate for Design of Museums during the major renovation of the museum and the reconstruction of the Hermes Hall. Several solutions combining sliders and/or dampers were investigated, and a system of four friction pendulum devices was selected. The main phases of the project were the design of the isolation system and strength verification of the statue, the purchase and testing of the bearings, and the construction and installation. In addition, instrumentation and monitoring of the response in situ and a full-scale experimental verification of a replica of the statue have been scheduled in the near future.

The design of the isolation system was based on a seismic hazard analysis that contained far-field and near-field excitations from the surrounding area, major seismic events from a broader area, and scaled strong ground excitations around the globe. The strength verification of the statue was performed by a detailed finite element analysis that considered the different material properties of the statue. The geometry of the model was produced by an accurate scanning technique that created a stereolithographic model. From this, a tetrahedral solid finite element model was created that was used in the analysis. The four friction pendulum system (FPS) bearings, provided by EPS, were tested at the State University of New York at Buffalo.

A large pit located below the museum gallery floor was constructed during the early building of the Hermes Hall in the renovated museum. This pit accommodated four large support pedestals. Four bearings were installed and leveled to withstand a 4 by 4 m square (approximately 13.2 × 13.2 ft.) composite plate on which the base and statue were anchored. The displacement capacity of the system is 32 cm (approximately 12.5 in.), while the maximum transmitted acceleration at the base of the statue is limited to 2.6 m/sec.^{2} (8.5 ft./sec.^{2}), which the statue can safely accommodate.

The Statue of Hermes

The statue of Hermes carrying Dionysos (fig. 1) is the only known surviving work of the famous sculptor Praxiteles and belongs to the Classical era (343 B.C.). It is made from a white Parian marble and has a height of 2.13 m

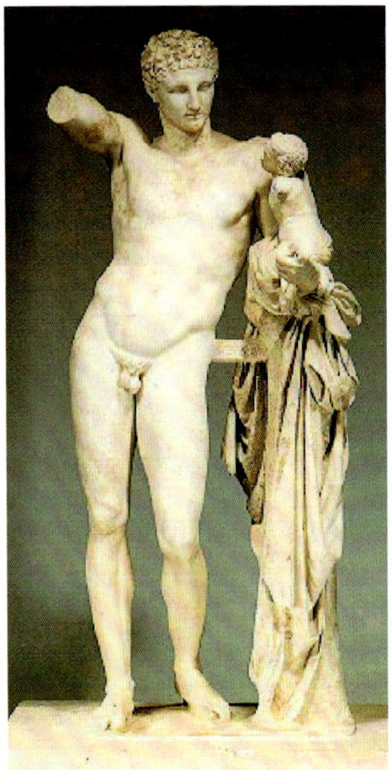

Figure 1
Praxiteles
Greek, active ca. 375–340 B.C.
Hermes Carrying the Infant Dionysos, 343 B.C.
Marble, H: 2.13 m (83¾ in.)
Archaeological Museum of Olympia, Greece

(about 7 ft.). It was found in 1877 in the Temple of Hera at Olympia. The right hand of Hermes, now missing, is thought to have held grapes or another object that held the attention of the crying Dionysos. Also missing are the left leg below the knee, the lower part of the right leg, and the base of the tree. All of these segments have been replaced by gypsum plaster restorations and supports. The statue is considered one of the most characteristic works of the fourth century B.C., a time when realism prevailed. The sculpture presents a number of different moods, depending upon the viewer's angle of view. Looking at the right side the statue, one has the impression that Hermes has a sad expression. If one looks at the left side, a smiling expression is noticeable. The frontal view, interestingly enough, suggests a more natural expression.

Design of the Isolation System

The basic requirement of earthquake-resistant design to protect this type of artifact is to prevent overturning and to provide adequate stability and strength in the case of strong ground motion. Conventional design methods aim at increasing the strength, which in the case of monuments—and especially artifacts—is prohibited or limited, since, for such a step, radically invasive techniques would be required. Alternatively, seismic isolation aims at reducing the seismic demand on the monument or artifact itself, so that the structure remains intact. Since earthquakes cannot be controlled, the only way to reduce seismic demand is to intervene between the structure and the ground with special devices that absorb the main part of the seismic energy and transfer only a small part to the superstructure or object (Naeim and Kelly 1999; Kelly 1996).

There are several applications of seismic isolation systems to artifacts all over the world—in countries such as Japan, the United States, and Italy. Well-known applications of seismic isolation to artifacts include *The Gates of Hell* by Rodin in the National Museum of Western Art in Tokyo and the statues of Scylla and Neptune in the Regional Museum of Messina, Italy. The J. Paul Getty Museum also has a long tradition of using its own two-degrees-of-freedom sliding device to isolate statues and smaller artifacts. In the case of the statue of Hermes (and its base) at the Archaeological Museum of Olympia, the layout described in figure 2 was proposed, after consideration of several alternatives for isolation systems that included designs incorporating Teflon and viscoelastic materials.

The statue and its base were scanned to develop a stereolithographic model (fig. 3). Parts made from materials other than marble were identified, as were the anchors and dowels that were used to assemble the constituent parts. The anchors and dowels were identified with a radiographic technique (figs. 4–6).

Finite Element Analysis

Based on the stereolithographic model of the outer surface of the statue and its base, a 3-D finite element model of the superstructure with tetrahedral elements was prepared, together with images of all anchors (fig. 7). In Table 1, the properties of the different materials used in the analysis are listed. The model consisted of 253,941 nodes, 158,303 tetrahedral elements, and 756,900 degrees of freedom.

The analysis of the isolation system was based on a 3-D Basis-ME program with some modifications (Katsaras, Koumousis, and Tsopelas 2004).

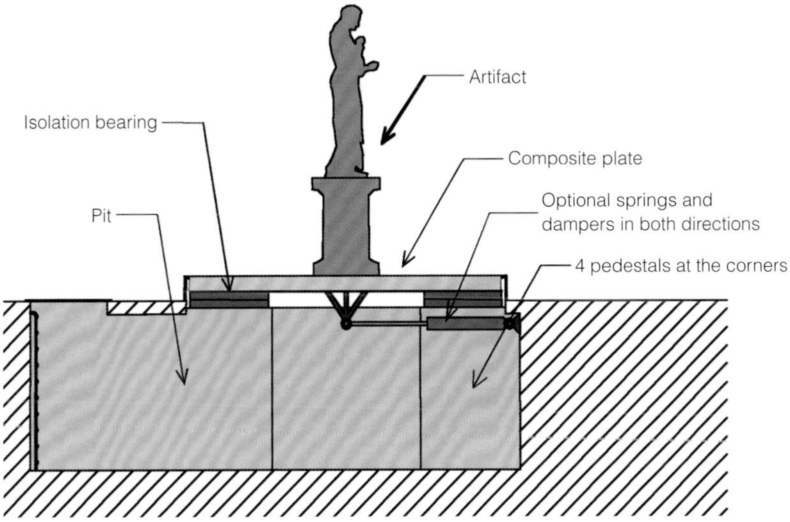

Figure 2
Layout of the base isolation system for the statue of Hermes.

This program requires as input the fundamental mode shapes of the superstructure and the three components of the strong motion record (fig. 8). As a result, the time history of the motion of the base is calculated at specific time steps.

The vertical load, *N*, is considered to vary with time at every bearing, due to the vertical component of acceleration and the overturning moment,

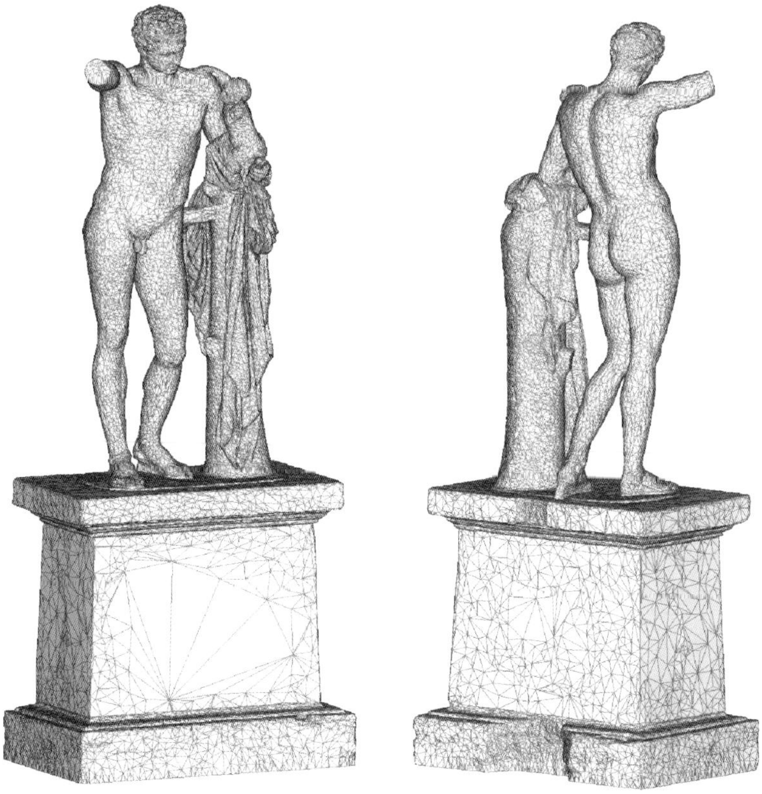

Figure 3
Stereolithographic model of the Hermes sculpture and base.

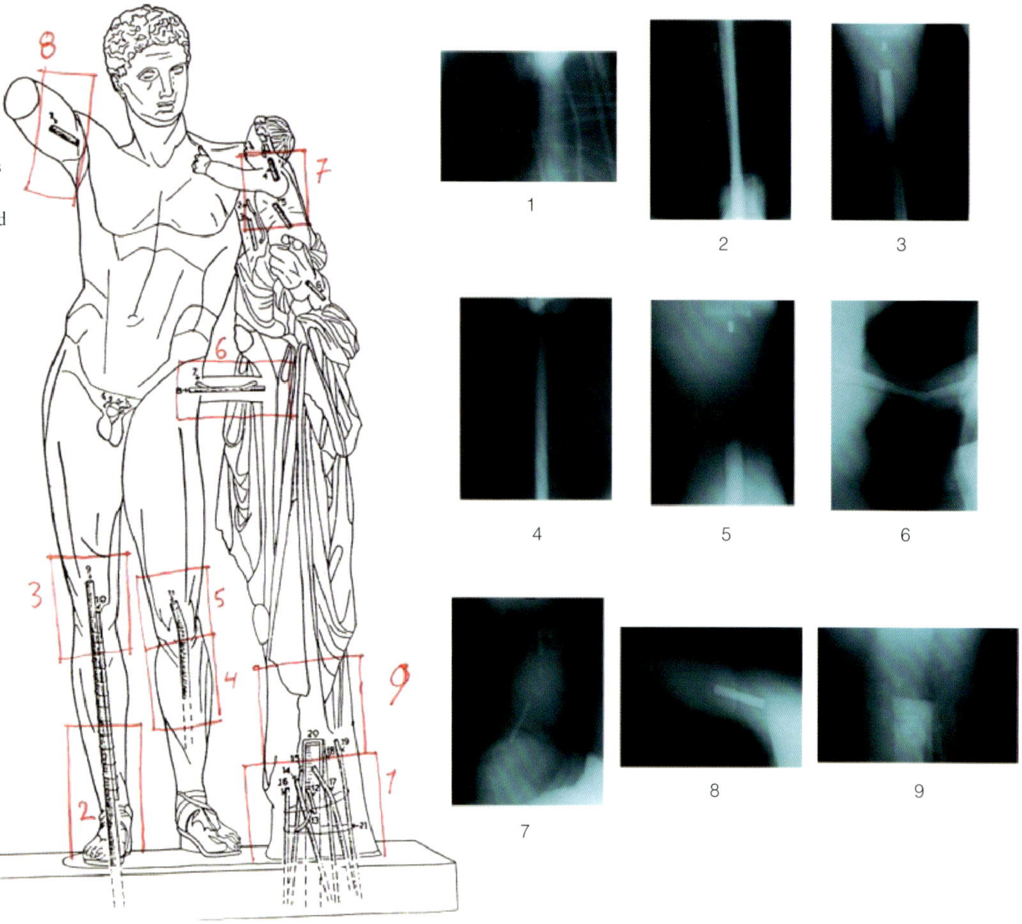

Figure 4
Radiographic images of the various metal connectors embedded in the sculpture.

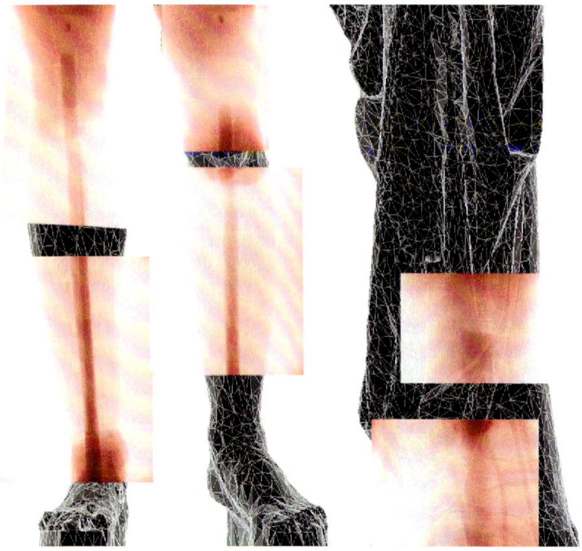

Figure 5
Detailed radiography of the lower part of the statue.

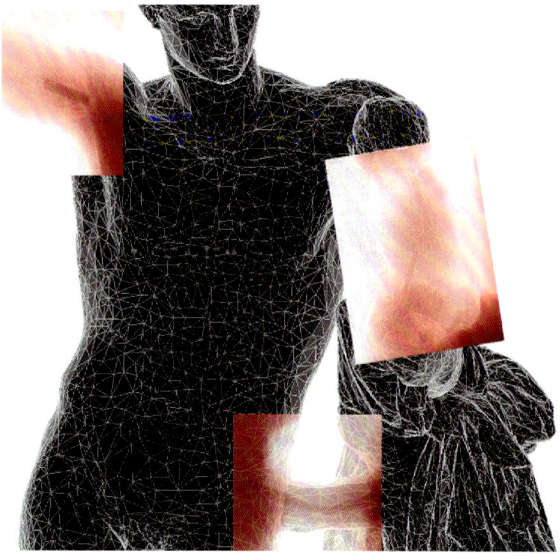

Figure 6
Detailed radiography of the upper part of the statue.

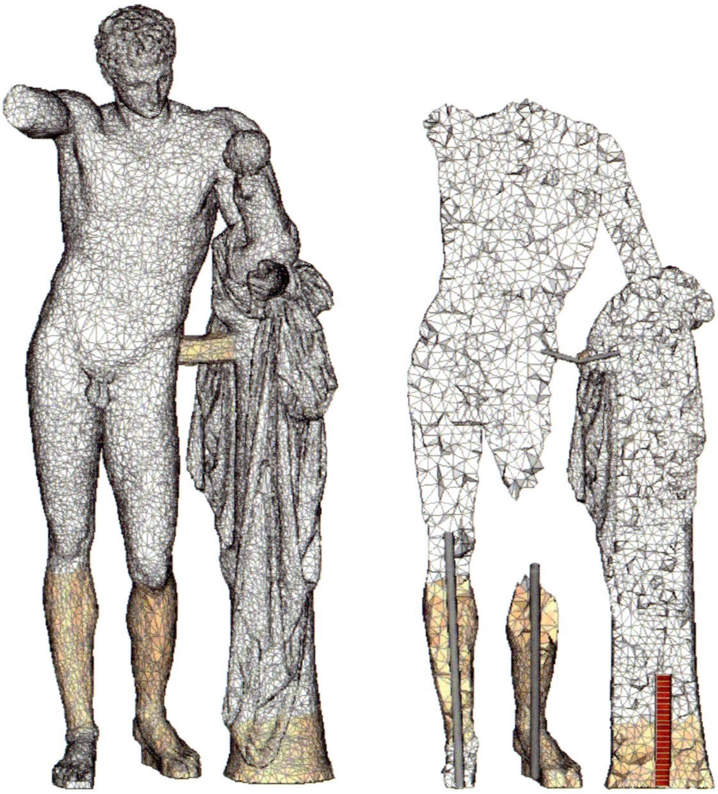

Figure 7
The 3-D finite element model showing connectors made of different materials (gray = bronze connectors; red = steel connectors).

Table 1
Properties of the materials composing the statue.

Properties	Marble	Gypsum plaster	Bronze	Iron
Density (kg/m^3)	2690	900	8300	7200
Modulus of elasticity (GPa)	47.0	5.0	108.0	126.5
Poisson's ratio	0.25	0.23	0.34	0.25
Compression strength (MPa)	33.0	7.8		
Tensile strength (MPa)	3.0	1.2		69.0
Yield stress in tension (MPa)			82.0	

Figure 8
Fundamental mode shapes of the statue.

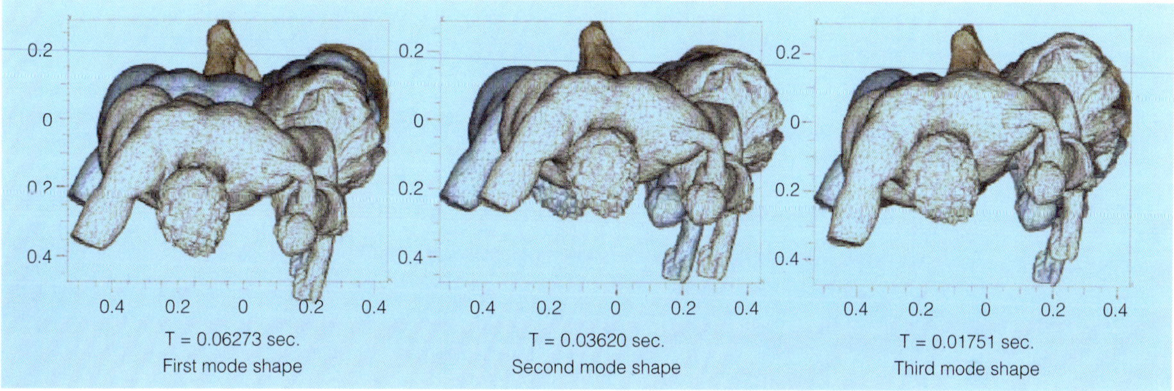

according to the following relation (in the equations that follow, measurements are given in metric units):

$$N = W_i \left(1 + \frac{\ddot{U}_V}{g} + \frac{N_{OM}}{W_i}\right)$$

where W_i is the initial weight on the bearing, $\ddot{U}_v$ is the vertical component of acceleration (positive upwards), and N_{OM} is the additional load due to the overturning moments (positive for compression).

Variations of the vertical load affect the friction force, which is a function of velocity and pressure. The coefficient of friction is related to the velocity by following the relations proposed by M. Constantinou (Constantinou, Mokha, and Reinhorn 1990).

The restoring force of the friction pendulum system (FPS) modeled on the basis of the Bouc-Wen model (Koumousis 2001) is given as:

$$F(t) = \frac{mg}{R}U + \mu_s mg \cdot z(t)$$

where the nondimensionalized hysteretic variable $z(t)$ satisfies the following nonlinear differential equation:

$$\dot{z}(t) = \frac{1}{u_y}\left[A - |z(t)|^n(\gamma\, sign(\dot{u}(t)z(t)) + \beta)\right]\dot{u}(t)$$

Therefore, the equation of motion of the single bearing is given as:

$$m\ddot{u}(t) + c\dot{u}(t) + F(t) = f(t)$$

where $f(t)$ represents the excitation force.

Seismic Hazard

According to the *International Building Code (IBC) 2000*, isolation systems should be designed for two different levels of earthquake excitations: the design basis earthquake (DBE), which corresponds to a probability of exceedance of 10% over 50 years (return period, 475 years), and the maximum considered earthquake (MCE), which corresponds to a probability of exceedance of 2% in 50 years (return period, 2475 years). A number of earthquakes from the surrounding area, as well as other major seismic events were scaled to meet these requirements. The SRSS spectra of these earthquakes are presented in figure 9.

Analysis Results

On the basis of the above analysis for the different systems considered, the following isolation system was selected, mainly on the availability of the bearings (Katsaras and Koumousis 2002): Four FPS bearings (see fig. 10) with maximum displacement capacity U = 32 cm; R = 2.2 m; with period T = 3.0 sec. The maximum transmitted acceleration to the base is: a = 2.6 m/sec^2 (8.5 ft./sec.^{2}).

For the selected system, the resulting movements along the x and y directions in plan do not exceed 25 cm (9.8 in.), as, for example, for the magnified Alkyonides Earthquake, presented in figures 11 and 12.

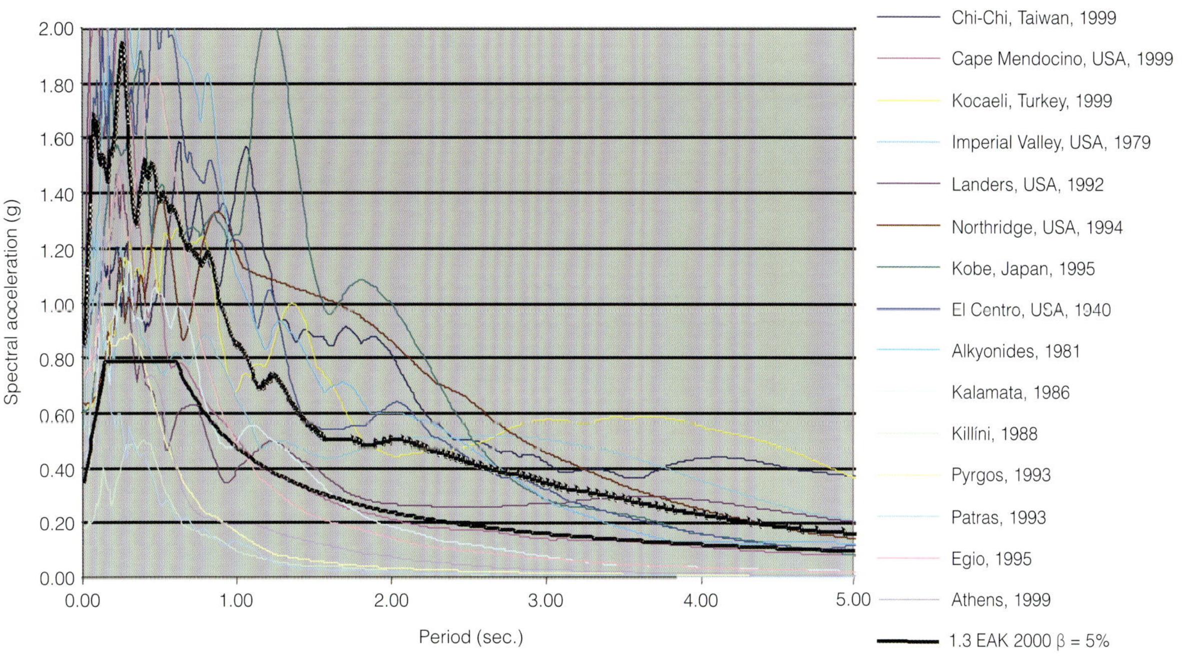

Figure 9
Earthquake design spectra.

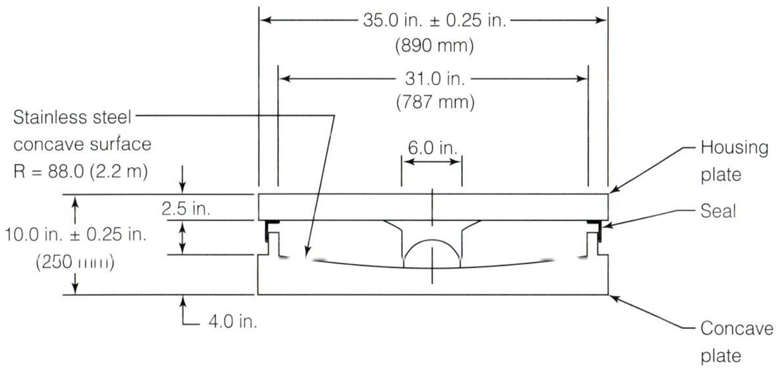

Figure 10
Section of friction pendulum system (FPS) with dimensions.

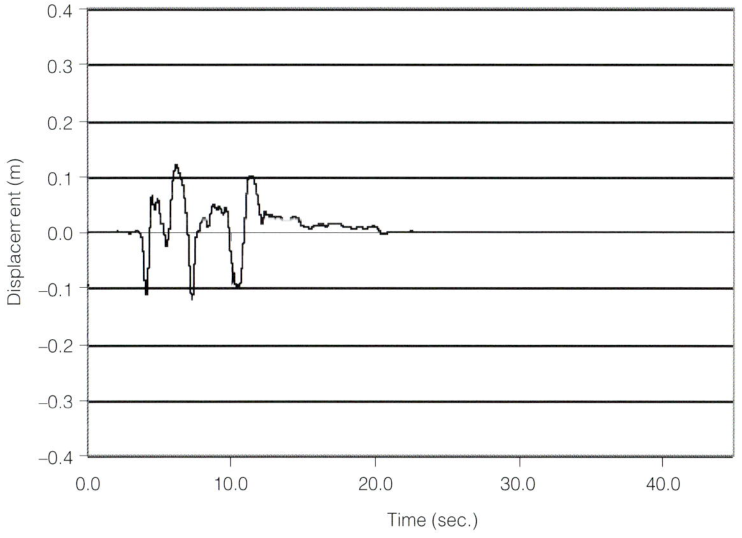

Figure 11
Displacement along the x direction for the Alkyonides Earthquake.

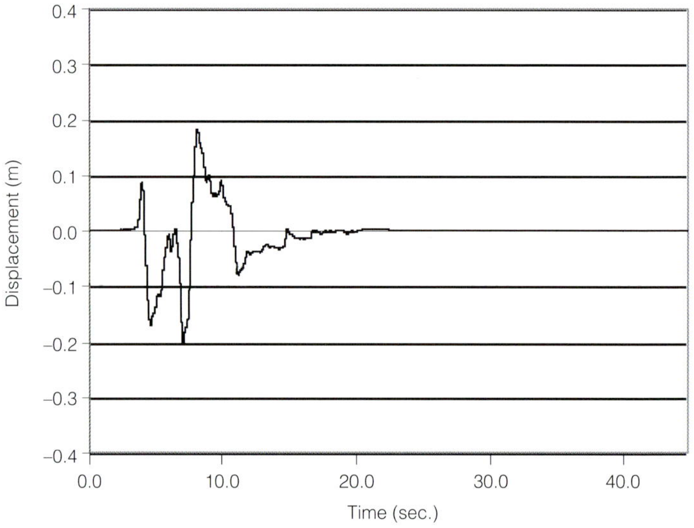

Figure 12
Displacement along the y direction for the Alkyonides Earthquake.

Stress Verification of the Statue

Based on the accelerations transmitted to the superstructure, a stress verification analysis was performed to validate the available strength of the statue, especially in the lower parts that are made from gypsum plaster, a material with significantly lower strength as compared to marble. In figure 13, the

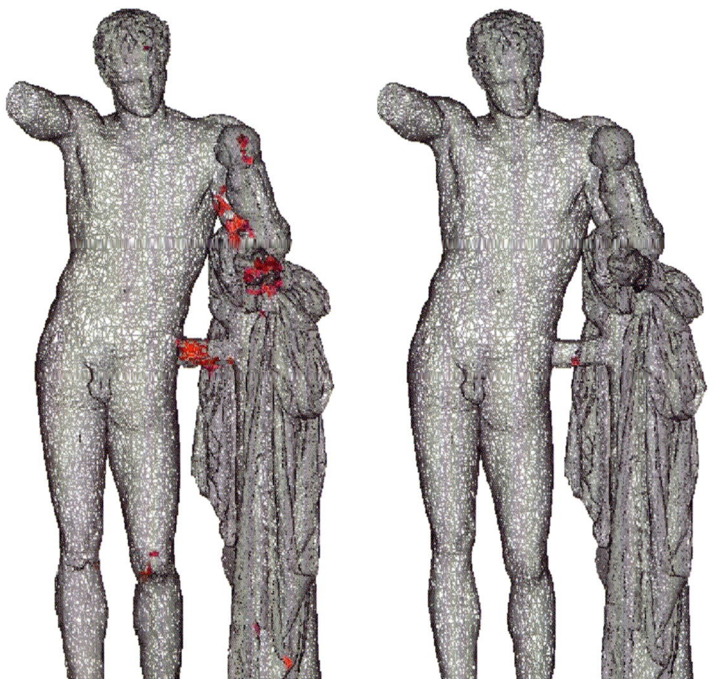

Figure 13
Overstressed areas (in red) for the nonisolated statue (a) and the isolated statue (b).

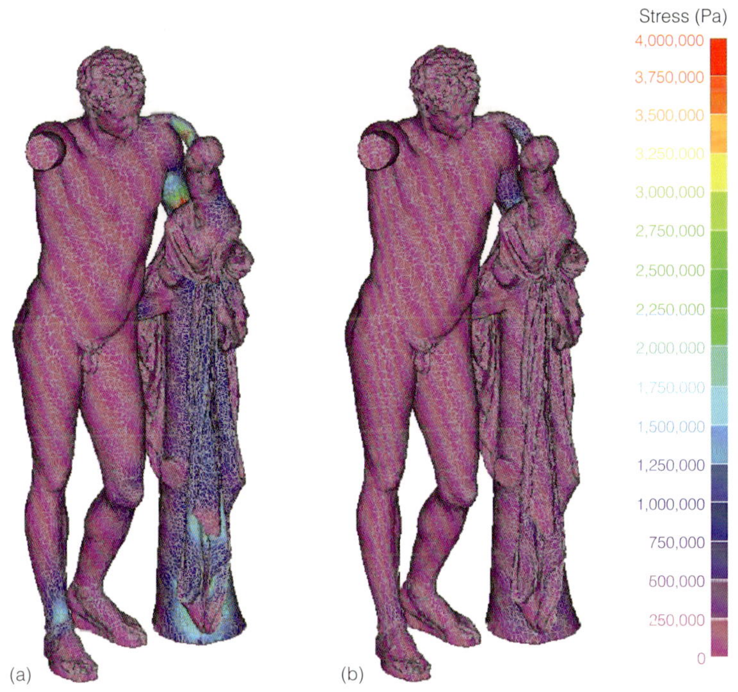

Figure 14
Major stresses on the nonisolated statue (a) and the isolated statue (b).

overstressed areas are designated for the nonisolated and isolated statue. This information reveals that the strength is adequate to sustain the severe seismic events considered in the analysis. Moreover, in figures 14 and 15a and 15b, the major and minor stresses in the statue are presented.

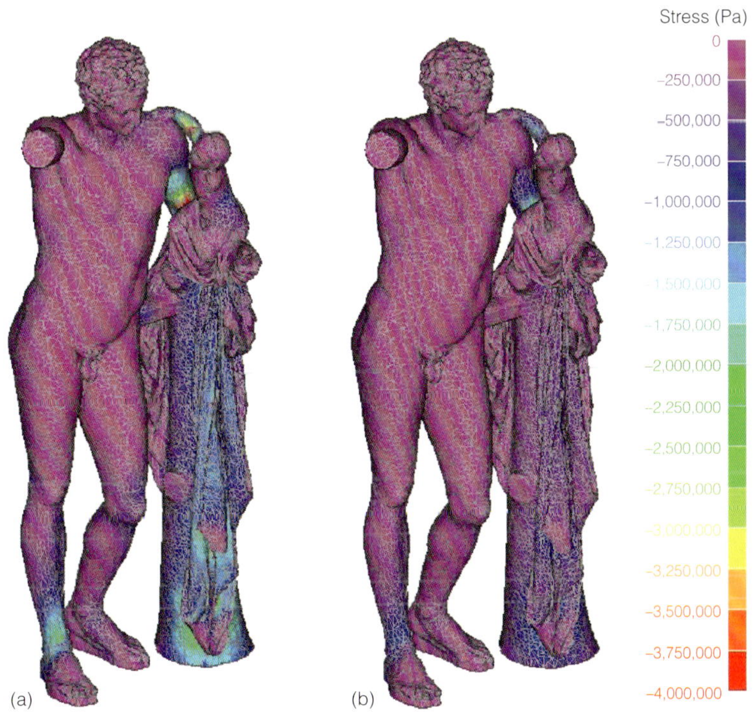

Figures 15a and 15b
Minor stresses on the nonisolated statue (a) and the isolated statue (b).

Conclusion

Seismic isolation represents an alternative strategy for the earthquake-resistant design of structures. It can be applied effectively for the protection of monuments and artifacts in cases where strengthening up to the level required by codes in seismic-prone areas is not feasible, as in the case of the statue of Hermes. Moreover, the relatively small weight of these structures (compared to buildings and bridges) imposes certain difficulties in selecting the appropriate reliable seismic isolation system. For the statue of Hermes, the FPS system proved adequate. The design of the isolation system was based mainly on the *IBC 2000* code.

Acknowledgment

The continuing support of the Ministry of Culture of Greece for the Hermes project is gratefully acknowledged.

References

Constantinou, M. C., A. Mokha, and A. M. Reinhorn
1990 "Teflon Bearings in Base Isolation II: Modeling," *Journal of Structural Engineering* 116, no. 2, pp. 455–74.

Katsaras, C., and V. K. Koumousis
2002 "Optimal Design of Lead Rubber Bearing Seismic Isolators in Buildings," *Proceedings of the Third International Conference on Engineering Computational Technology* (Stirling, UK).

Katsaras, C. P., V. K. Koumousis, and P. Tsopelas
2004 "Geometrically Nonlinear Spring and Dash-pot Elements in Base Isolation Systems," *Proceedings of the Ninth International Conference on Civil and Structural Engineering Computing*, ed. B. H. V. Topping (Stirling, UK).

Kelly, J. M.
1996 *Earthquake-Resistant Design with Rubber*, 2nd ed. (London).

Koumousis, V. K.
2001 "Non-Linear Dynamic Behaviour of Base-Isolators," *Proceedings of the Third Hellenic Society of Theoretical and Applied Mechanics Conference* (Thessaloníki).

Naeim, F., and J. M. Kelly
1999 *Design of Seismic Isolated Structures: From Theory to Practice* (New York).

Earthquake Mitigation at the Seattle Art Museum

Nicholas Dorman

Introduction

Seventy-four years have elapsed between the 1933 opening of the original Seattle Art Museum (now the Seattle Asian Art Museum) in Seattle's Volunteer Park and the imminent opening of the newly expanded downtown museum. Since the first Seattle Art Museum was established, the city of Seattle, the institution and the collections housed within the museum's multiple sites have grown enormously.[1] During that time span, the science of seismology and, in particular, the understanding of the seismic profile of the Puget Sound region have developed to such an extent that we can examine evidence of historic and even prehistoric seismic activity.[2] We are now able to predict the possible character and effects, although not the specific timing, of future earthquakes with increasing levels of precision.

This paper presents an account of the various seismic risks that communities in the Pacific Northwest region are subject to. It is also an account of the history of measures used at the Seattle Art Museum to mitigate the damage that earthquakes might cause to art on display and in storage.

Seismic Profile of the Pacific Northwest

The Puget Sound region is currently believed to be at risk from three primary types of earthquake, each of which threatens the region with distinct effects. These categories of earthquake include seismic activity occurring in the interplate region on the subduction zone off the coast, intraplate earthquakes originating many miles beneath the surface of the earth, and earthquakes on faults in the Puget Sound area. Not only are scientists able to define seismic activity in the Pacific Northwest according to these three broad categories of earthquake, they are also acquiring increasingly nuanced data pertaining to the precise geological character of the region. In addition to revealing the presence of formerly unknown crustal faults, these studies enhance our ability to predict the occurrence and results of landslides, ground liquefaction, and basin effects, all of which can amplify or compound the harmful characteristics of a seismic event.

Interplate Earthquakes: The Cascadia Subduction Zone

In the Pacific Ocean, a long plate boundary between the Juan de Fuca and North American plates runs in a southerly direction from the Brooks Peninsula on Vancouver Island to Northern California. Along this threshold, known as the Cascadia Subduction Zone, the heavier ocean plate plunges beneath the continental plate (fig. 1). Although this particular system has been

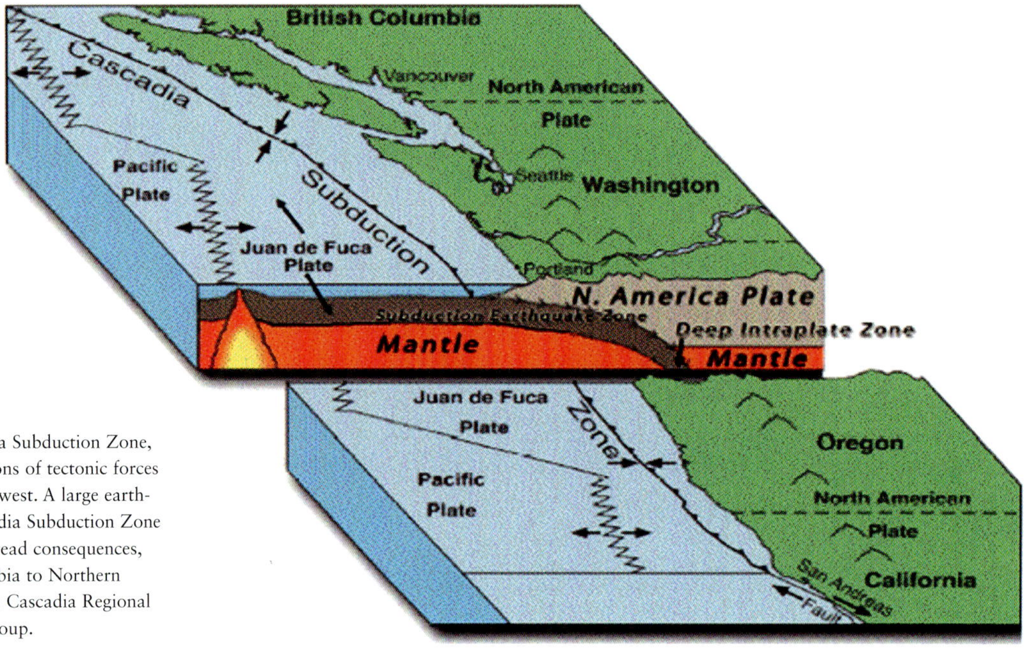

Figure 1
Map of the Cascadia Subduction Zone, showing the directions of tectonic forces in the Pacific Northwest. A large earthquake on the Cascadia Subduction Zone would have widespread consequences, from British Columbia to Northern California. Graphic: Cascadia Regional Earthquake Workgroup.

free of seismic activity in recent times, it is known that earthquakes of significant magnitude do occur on such plate thresholds. Such earthquakes result when the rocks on the two sliding plates lock, deform, and retain enormous stresses, eventually springing back into place. While the last interplate earthquake on the Cascadia Subduction Zone occurred in 1700 (Atwater et al. 1995), the catastrophic effects of this type of seismic event were recently seen in Sumatra. There, a large subduction earthquake and subsequent tsunami wreaked havoc on the lands bordering a sizable portion of the Indian Ocean.[3]

With a subduction earthquake, there is a preponderance of long waves of energy released over several minutes. Since this type of earthquake is only known in this region from historical evidence, we currently have little more than projections of the effects that might be caused. We have no precise indication of the way that tall and long structures in Puget Sound communities will behave when exposed to long wavelengths of energy of significant magnitude for a prolonged period, although the damage that was sustained to Mexico City in the 1985 earthquake (magnitude 8.1 on the Richter scale) might be taken to indicate the potential significant and devastating effects.[4]

If an earthquake were to occur along the entire length of the Cascadia Subduction Zone, it might reach magnitude 9 on the Richter scale at the event site just offshore (the magnitude might possibly be somewhat lower inland). The earthquake could last 5–7 min. and would cause regional damage along the coast. If the pressure along the subduction zone were to release in sections, each earthquake along the fault might reach an intensity of magnitude 8.5 and last 3–4 min.[5]

Recent scenario studies have taught us that, in addition to the trauma and damage caused by the initial earthquake, the adverse effects of a subduction earthquake will become compounded as time passes, owing to the region-wide character of damage occurring along the coast. An earthquake such as this may be expected to close ports and significant parts of the transportation and utilities infrastructure—damage that will have consequences for the economy and for society in the region. Understanding of these factors is, at first glance, discouraging. But knowledge of the threats facing us can help— indeed, is essential—in our development of emergency response strategies. Hurricane Katrina provided clear evidence of the ways in which a regional disaster isolates communities and institutions, disabling existing services and impeding potential avenues of relief.[6]

An earthquake on the Cascadia Subduction Zone would be likely to generate a tsunami that might quickly cause damage to the Northern California, Oregon, Washington, and British Columbia coasts and, potentially, to communities in Alaska and across the Pacific Ocean.[7] We might suppose the effect of this primary tsunami to be lessened and for it to dissipate somewhat after it has traversed the Straits of Juan de Fuca and entered Puget Sound, but any sense of relief in communities in south and central Puget Sound might be confounded by local conditions, interference effects, and other variables. Significantly for Seattle, local wave surges or seiches could be caused by landslides—occurrences that are typical for seismic activity in the region. Tectonic movement that results in a tsunami can also be the direct cause of a change in ground level that can compound the catastrophic influence of the inundation.

Aftershocks of significant magnitude are considered to be characteristic for subduction earthquakes. The magnitude of aftershocks could reach Richter magnitude 7, a level that exceeds that of the 2001 Seattle (Nisqually) Earthquake, although that earthquake belongs to a different category of quake, known as a Benioff Zone earthquake.

There is currently considered to be a 15% likelihood of a subduction earthquake occurring in the next 50 years.[8]

Intraplate Earthquakes: The Benioff Zone

This is the type of deep earthquake that is most familiar to people in the Seattle area from recent seismic activity. This variety of intraplate earthquake was the cause of the Richter magnitude 6.7 Nisqually Earthquake that hit Seattle and the surrounding vicinity in 2001. In the Pacific Northwest region, Benioff Zone earthquakes originate in the submerging Juan de Fuca Plate as it arcs below the North American Plate. Ruptures occur within the subducting plate as it reaches an angle of approximately 30°, releasing shock waves of medium-intensity energy. In the Seattle region, the Juan de Fuca Plate submerges under the overlying plate just off the coast. It reaches this critical submergence angle many miles directly below the communities of Puget Sound. In the case of the 2001 Nisqually Earthquake, shaking was felt for approximately 35 sec., and significant damage was sustained, notably to unreinforced masonry structures in the Pioneer Square historic district and to the Alaskan Way highway viaduct, which runs along the Seattle waterfront.

There is currently believed to be an 80% chance of this type of earthquake occurring in the next 50 years.[9] Deep earthquakes typically last for about a minute and are followed by few or no aftershocks.

Because this type of earthquake has occurred approximately once every decade, a reasonable amount of data has been gathered about the likely magnitude and consequences of a Benioff Zone deep earthquake. This information and experience serve a valuable function in the development of safety precautions for future earthquakes. There remains, however, a risk that familiarity with this category of earthquake might engender a sense of complacency that in no way prepares the population for the prolonged shaking of a subduction quake or the intense peak ground acceleration levels of an earthquake on a near-surface fault. In the absence, in living memory, of direct experience of these latter types of earthquakes, experts have initiated a number of scenario case studies to improve the region's threat recognition and emergency response capacity.[10]

Near-Surface, or Crustal Earthquakes

The Seattle area is vulnerable to crustal earthquakes, such as those that can occur on the Seattle Fault and myriad other surface faults in the region (fig. 2). In contrast to a subduction event, this type of earthquake is characterized by much shorter wavelengths of energy and high levels of peak ground acceleration. On the ground, a violent earthquake is experienced, with significant quantities of debris being shaken from structures. Building code is designed to maximize life safety in structures that are subjected to earthquakes on a surface fault. It is believed that 80% of the damage sustained in a crustal earthquake would occur as nonstructural building components and contents of buildings become detached or are shaken from their usual positions. Because of the number of buildings that fall short of current code standards, however, it is anticipated that a crustal quake would close the city, at least for a short period.

There are similar surface faults in the Puget Sound area on Whidbey Island and elsewhere, but the primary fault of concern for the city, and for the museum, is the Seattle Fault, which traverses built-up areas only a short distance from the downtown museum location. This fault has been the object of much recent research.

There is approximately a 5% chance that a crustal earthquake will occur on a surface fault in the next 50 years. The magnitude of this earthquake might be as much as 6.8 on the Richter scale, and sizable aftershocks are to be expected.[11]

Other Threats

These three types of earthquakes—subduction, Benioff Zone, and near-surface—are the primary causes of seismic concern for staff at the Seattle Art Museum, but as we manage risks to the collection and plan for emergency response, we also consider other threats and develop procedures to reduce their

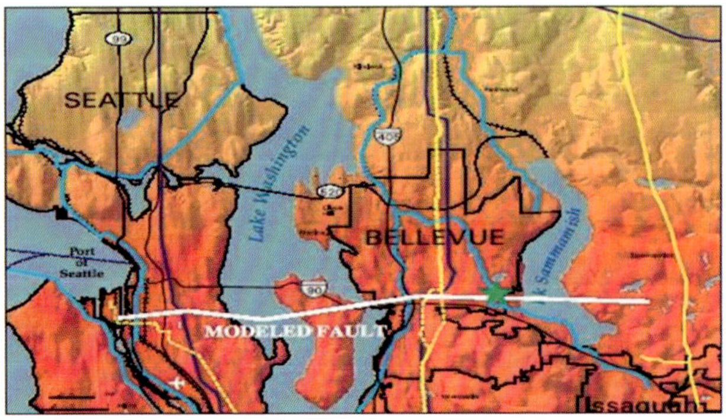

Figure 2
A surface fault in Seattle. In a recent scenario study, the Earthquake Engineering Research Institute and the Washington Military Department Emergency Management Division (2005) analyzed the impact of a Richter magnitude 6.7 earthquake on a fault in the earth's crust. The scenario was based on a known fault, one of several located in densely populated areas in the region. Graphic: U.S. Geological Survey.

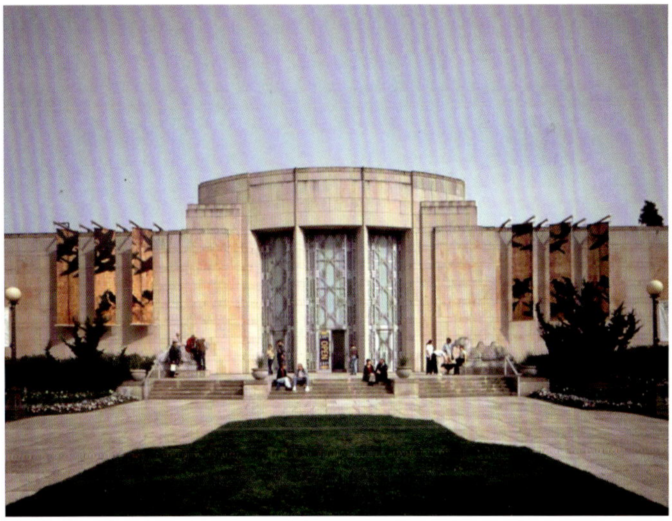

Figure 3
The Seattle Asian Art Museum in Volunteer Park. This building held the original Seattle Art Museum, which opened to the public in 1933. The historic structure has been renovated for improved seismic stability and life safety. Photo: Seattle Art Museum.

harmful effects. These threats include volcanic ash fall and tsunamis, as well as secondary effects from an earthquake, such as fire or flooding.[12]

Seismic Issues at the Seattle Asian Art Museum

The Seattle Asian Art Museum (SAAM) in Volunteer Park, designed by architect Carl F. Gould, opened in June 1933, a concrete Art Moderne building with a facade of Wilkeson sandstone (fig. 3). On the interior of the museum, extensive use was made of hollow clay tile for non-load-bearing walls.

The building served as the original Seattle Art Museum until the construction of the downtown building in 1991. From the institution's inception, an arrangement was made that the structure of the Volunteer Park museum building would be maintained by the City of Seattle, and the art program would be run by the museum (Fuller 1993).

Although the building and location had traditionally been considered relatively seismically stable, consideration of earthquake safety matters in the 1990s brought an increased awareness on the part of the museum and the city authorities. They realized that a number of seismic issues needed to be addressed in order to bring the museum in line with construction code standards. From 1993 to 1994, work was undertaken to clean and reattach the top course and the capstones of the facade in accordance with the then-current *Uniform Building Code*. Brick-and-mortar portions of the entry-level wall were replaced with reinforced concrete, and the facade was reattached. Lintel stones at the entrance were bolted to the structure with reinforcing steel and epoxy resin.[13]

In 1995 Leavengood Architects was commissioned by the City of Seattle Department of Construction and Land Use to determine building occupancy loads, emergency exit routes, and structural problems that might threaten life safety at the exits during an earthquake (Leavengood Architects 1995). This study revealed further problems owing to the prevalence of the hollow clay tile walls and to poor connections between the original structure of the building and extensions, which had been added in 1953 and 1955. Deficiencies were also noted in the fastening of horizontal diaphragm structures to shear walls in these extensions. Many of the problems presented by the hollow tile walls in public areas have been addressed over the years, and inadequate connections in the Alvord Boardroom have been reinforced with a steel and epoxy-injected bolting system. Two of the hollow clay tile walls in the auditorium have been demolished and replaced by metal stud and sheetrock/plaster walls; a third wall has been reinforced with a metal grid system tie-backing the wall to the structural elements.

In 2004, the 2003 edition of the *International Building Code (IBC)* was adopted in Seattle, and it superseded the various iterations of the *Uniform Building Code* that had gone before. The new *IBC* further refined and tightened the guidelines for seismic safety, particularly in relation to the fastening of nonstructural components, and the most recent seismic upgrades have been implemented in accordance with *IBC* standards. During these renovations, the process of reconstituting effective diaphragms in the museum has been continued. The connections between the shear walls and the steel girders that hold the exterior skylights have been augmented with structural steel and an epoxy-injected bolting system. Tubular steel attachments in the attic have been used to connect open diaphragms over the skylights to the concrete walls on either side, and original concrete roof sections have been fastened more firmly to the walls by reinforcing the attachment every few feet with angle-iron L brackets and the specified epoxy-injected bolting system.

By these incremental measures, a gradual reduction of the seismic vulnerability at the Seattle Asian Art Museum has been achieved in an affordable manner. The 1933 building might be considered relatively secure, as a low-lying structure on firm ground, during a subduction earthquake, and the renovations of recent years will most likely prevent loss of life in a severe quake on a surface fault. The next necessary step in the protection of the collections, however, is the precise characterization of the level of threat to works of art housed in the galleries and storage facilities. During a subduction earthquake, long and tall structures within the museum would remain particularly vulnerable, unless they were sufficiently well anchored. We also need to bear in mind the fact that recent seismic upgrades removed or secured hollow-core tiles only in areas where public safety might be jeopardized. No improvements were made to walls adjacent to artworks in storage, although these and other measures are currently being considered as part of the latest proposed round of seismic improvements.

Seismic Issues at the Downtown Seattle Art Museum

In 1991 the Seattle Art Museum (SAM) opened a new facility in downtown Seattle (fig. 4). The building, designed by Venturi, Scott Brown and Associates, provided the institution with a modern museum structure with increased exhibition area and extensive exhibition support facilities. With the opening of the new building, the original museum became the Seattle Asian Art Museum, and the collections on display and in storage were divided between the two museums.

Venturi's building was designed to provide excellent rigidity along the east-west axis in line with the slope down toward the Seattle waterfront. This rigidity was achieved by means of a substantial shear wall[14] along the north side of the structure and by further rigid walls around the stair and elevator shafts.

The Venturi building was essentially a first step in the development of the Seattle Art Museum presence in the heart of downtown Seattle. When viewed

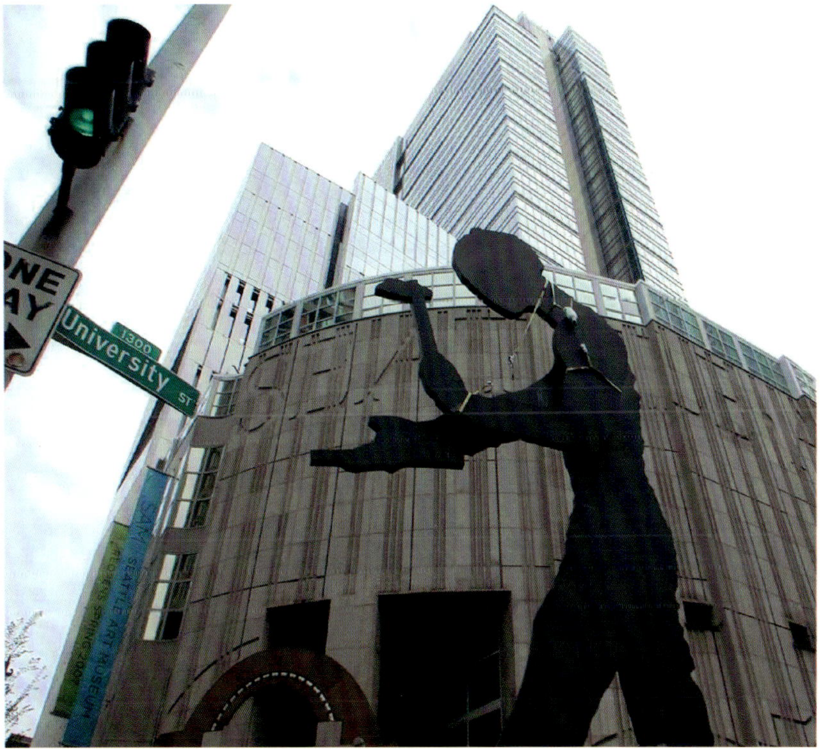

Figure 4
The 1991 downtown Seattle Art Museum (foreground) and the new adjoining tower (background), which houses the Seattle Art Museum North Building and the Washington Mutual Bank headquarters. Photo: Nicholas Dorman.

from the exterior, the "unfinished" appearance of the flat concrete north wall made no attempt to disguise the fact that the building would, one day, become a fraction of a much larger structure, filling the empty lot to the north. As the museum moved into the twenty-first century, the tentative plan to demolish the adjacent car parking facilities and a seismically unstable former department store and use these areas for an expanded museum remained the goal. A proposal from Washington Mutual Bank in 2001 was not, however, anticipated. The construction of a forty-two-story bank tower adjacent to the Venturi museum building, with an attached expansion for Seattle Art Museum galleries, was now being considered. At present, the bank offices have already been occupied, and in April 2007, four new floors of museum galleries, public space, and retail and restaurant facilities will open. The SAM premises will eventually expand further to encompass twelve floors of museum facilities as upper levels that are temporarily rented to the bank are taken over by the Seattle Art Museum.[15]

What has seemed a highly opportune venture for the museum also presents a number of serious issues. In addition to the complications arising from sharing joint occupancy within a common building, there are challenges associated with maintaining a stable environment in a vertical structure with a relatively open floor plan. And there are seismic problems that need to be resolved. How would the buildings best be designed and connected to overcome the inevitable structural asymmetry and to ensure adequate stability performance in an earthquake? What might be the consequences for objects on display or in storage in a vertical building?

Despite the inherent rigidity of the Venturi museum structure, engineers for the expansion/tower project were concerned that an earthquake might result in destructive pounding between the old and new buildings. To reduce or eliminate the effect of pounding, the two structures were connected with 1000 metal ties. While these ties bind the two structures tightly together along the north-south axis, the Teflon-coated surface and the shape of the steel attachments permits some slippage movement along the east-west axis, which, as described above, is quite rigid in the case of the Venturi building.[16] The new tower has also been engineered to be relatively stable for a high-rise structure, a characteristic that reduces the potential for each building to behave differently during an earthquake.

In addition to the ties binding the old and new buildings together, the relative difference in rigidity between the two structures was further reduced by cutting a number of thoroughfare holes in the thick concrete north wall of the Venturi building. These holes provide routes between the two buildings.

Prior to the construction of the tower, support piles were augured into the ground below the site and anchored deep into the glacial till that lies below the surface in this part of Seattle. The foundations were stabilized further with a 14 ft. thick mat of reinforced concrete.

The seismic stability of the tower derives from a central box, composed of a reinforced concrete services core, buckling-restrained braces, and tubular, concrete-filled pipe columns with cross bracing. The design was complicated further by the manner in which the building steps back, reaching a total of sixteen stories in height on the west (museum) side and forty-two stories on the east (bank) side (fig. 5).

The new building has been built to rigorous design standards, and the proposal was subject to peer review from engineers in California. The performance targets for this building are for life safety in a 500 year quake.[17] In fact, the engineers are confident that the design will provide greater performance than would be required for the provision of life safety alone. For the museum, however, additional issues will need to be addressed to discover more about the new building: how, exactly, will the SAM-occupied floors behave during earthquakes of various magnitudes? What differences will there be between the seismic performances of the old and new buildings and between upper and lower floors? Clearly the design criteria enshrined in the building code need to be tailored to guarantee life safety as far as possible in an earthquake—but how much further do we need to go to protect the museum's irreplaceable collections and to forestall the need for extensive repairs following a seismic event?

When faced with the prospect of joint occupancy of a common structure, one issue of considerable importance that the museum had to confront was the question of water containment. Familiar with containment issues from the sloping site of the existing downtown museum, SAM staff demanded stringent measures to preclude the possibility that water leaks originating high in the tower would damage art on the floors below. The risk of leaks occurring during an earthquake was addressed by installing a waterproof membrane across the entire west half of the building on the twelfth floor, which will eventually be occupied by SAM offices. This strategy gave the museum control over penetrations and provided a "buffer" story, should water come from above. In addition to this precaution, restrooms and all other plumbed facilities were grouped to the east side of the building, with floors that are sloped to drain into encapsulated pipes. A dry-pipe preaction sprinkler system is used in and above the galleries.

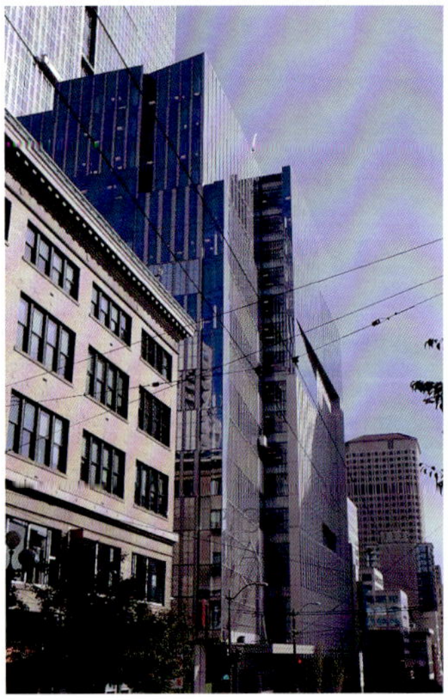

Figure 5
The Seattle Art Museum North Building and Washington Mutual Tower, viewed from the north. SAM will eventually fill twelve floors of galleries and support space in the lower section of the North Building (center). Photo: Susan Lewandowski.

In the expanded museum galleries, new installation challenges requiring particular consideration of seismic issues included the installation of a centuries-old Italian paneled room. The antique panels were attached to metal stud supports with input from engineers who determined the requisite number of attachment points. The aim of the installation was to fasten the room in place adequately, while satisfying the conservation requirement of penetrating the wood only as much as was absolutely necessary.[18]

Another challenge was the display of a thousand pieces of porcelain in one gallery—each piece was provided with a secure, custom-made mount by the professional mount making staff at SAM. These mounts need to be extremely secure, owing to the large number of porcelain pieces attached, one above another, directly to the wall on individual gilded brackets. Onerous in a different sense, yet also requiring consideration of seismic issues, was the placement of nine cars high above the heads of visitors in the public "Forum" space of the new building, for Cai Guo-Qiang's installation *Inopportune: Stage One*.[19]

Mount Making at the Seattle Art Museum

SAM is fortunate to have well-developed mount making facilities and expertise. The 1980s were marked by a paucity of published material relating to the mitigation of seismic effects through appropriate mounts. SAM mount makers conducted extensive research into the development of mounts for the protection of the collections, and, as a part of this program, they traveled to the J. Paul Getty Museum, where they worked with mount maker Wayne Haak. They returned to Seattle with ideas and practices that were then applied, with adaptations to suit the extraordinary variety of objects in the SAM collections. The urgent need for seismically designed and effective mount making became apparent in the days after the Loma Prieta Earthquake of October 17, 1989, in which millions of dollars worth of damage to art in San Francisco museums was sustained.

SAM mount makers extrapolated principles for stabilizing art in the galleries from their own experiences and from studies undertaken at the Getty Museum,[20] and they applied these principles, within the constraints of budget resources, to the care of collections at SAM/SAAM (figs. 6 and 7).

In the following years, SAM mount makers developed a comprehensive repertoire of approaches to making seismically stable mounts. Static mounts, employing full capture of objects, were developed to provide effective support for works of art. These mounts obviated any need for spring-loaded assemblies. They were designed to hold a work of art securely during an earthquake and to stop it from chattering or falling from the mount.[21]

By fastening art with fixed mounts, the need for supplementary stabilization (such as lead-shot bags placed inside vessels to secure them) could be reduced,

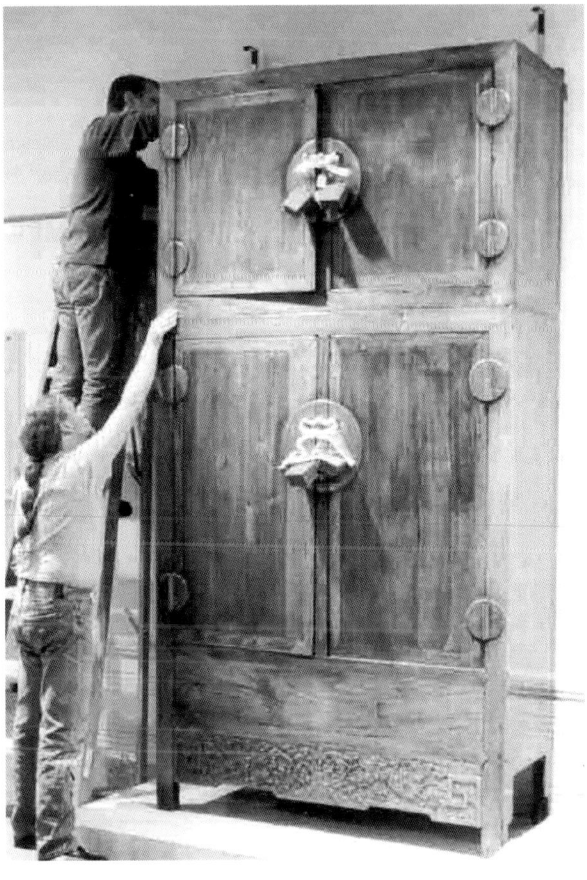

Figure 6
Mount makers securing an object in the galleries at the Seattle Art Museum. Furniture is fastened to the wall with padded, telescoping steel brackets. Photo: Gordon Lambert.

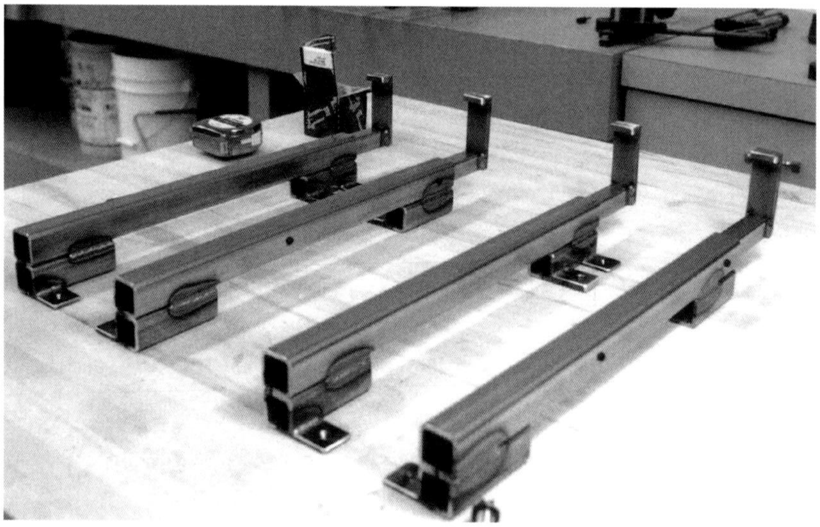

Figure 7
Mounts designed and produced at SAM, which were used to secure the cabinet shown in figure 6. Photo: Gordon Lambert.

although techniques such as fixing art to a surface with wax continue to be used in certain situations. A fixed mount is usually attached directly to the exhibition case, which, in turn, is fastened to the wall or stabilized with the correct amount of ballast. In addition to these fixed mounts, SAM staff also employ principles of damping and partial isolation, using elastomeric materials to cushion art, particularly in storage, to protect it from shock and vibration.

Another vital prerequisite for critiquing and developing improved mount performance is documentation. SAM mount makers have been rigorous in maintaining thorough documentation of mounts. This documentation includes photographs and records of mounts and installations, preparation of mount books for loan exhibitions, and records of complex treatments (such as the documentation of the mount program for the stone statuary in the Garden Court at the Seattle Asian Art Museum). Another important aspect of this documentation is the recording of mount and case performance following seismic activity. Since 1991 there have been six earthquakes over a Richter scale magnitude of 5.2. In that time, cases were found to have moved less than 9.5 mm (3⁄8 in.) following a magnitude 5.6 earthquake in May 1996, and only one object broke free from its mount during the 2001 magnitude 6.8 Nisqually Earthquake (although this incident was determined to be the result of an installation error and not caused by improper mount design).[22]

Although, to our knowledge, only one base isolator has been installed on a particularly valuable and vulnerable sculpture in Seattle,[23] it is hoped that further studies of isolator applications can be pursued. SAM conservation staff and mount makers would like to work with colleagues at other institutions to examine the performance of isolators under various circumstances corresponding to different categories of potential earthquakes in the various Seattle Art Museum environments.

Monumental Sculpture

For some works of art in the Seattle Art Museum collection, seismic stabilization is a life safety issue as much as preservation issue. To this category of art belong the large Chinese stone tomb statuary figures that now grace the grand stair of the Venturi building. The securing of the tomb guardian, camel, and ram figures was another early example of collaboration between SAM staff and colleagues at the J. Paul Getty Museum and in the private sector. SAM's then–deputy director, Gail Joice, worked with conservator Jerry Podany of the Getty Museum and Patricia Leavengood (then Tuttle), a conservator in private practice in Seattle, as well as structural engineers Leslie E. Robertson Associates (Leslie E. Robertson Associates 1989) and the University of Washington Department of Civil Engineering, to determine the risks associated with the proposed location of the art. The team also sought to devise a means of protecting the public and the art in that location.

In June 1989 Jerry Podany wrote a prescient report that preceded the Loma Prieta Earthquake by only a few months (Podany 1989). In this document he articulated the consequences of neglect of seismic issues and the need for museums to invest professional resources to protect their collections from the risks associated with natural disasters such as earthquakes. Also addressed was the gap between the provision of code guidelines for life safety and the general dearth of consideration for the impact of earthquakes upon collections housed within museums in vulnerable areas.

A model for an approach to the seismic care of collections was set out: the collections administrator must understand the seismic environment, understand the building structure, understand the object, and apply appropriate protection for the object.

In this instance, a seismic isolator for each sculpture could not be funded, but the studies resulted in the correct orientation of the art for maximum safety, the lowering of the center of gravity of each sculpture by tying the structure together and the provision of an appropriately sized and strengthened plinth and attachments, as well as local reinforcement of the staircase.

One factor that has greatly enhanced the museum's ability to partake in effective collaborative projects for the benefit of the collections is the establishment of the museum's first in-house conservation department and studio in 2001.[24] The conservation department works closely with staff in the mount making and other SAM departments and with colleagues in other organizations to institute treatments, risk management, and museum policy pertaining to seismic issues.

Storage

In art storage, seismic issues must also be considered carefully. While on display, each work of art is given a custom-made mount that protects it from theft or earthquakes. The vast quantities of art in storage, however, must be protected in a different fashion. When artworks are on display, security and aesthetics both dictate the final choice of mount. In storage, security and access are of primary importance. Fragile objects are protected in secure cabinets with foam drawer liners and partitions (fig. 8). The aim of these partitions is to afford the objects partial isolation from the forces of seismic activity and to protect them from other objects

The storage units are secured to the floor or to a wall, and in some places, freestanding units have been secured to one another for additional stability. All shelves and stacking units are anchored in place. Drawers are secured behind locking doors or secured from the outside with steel rods. All open

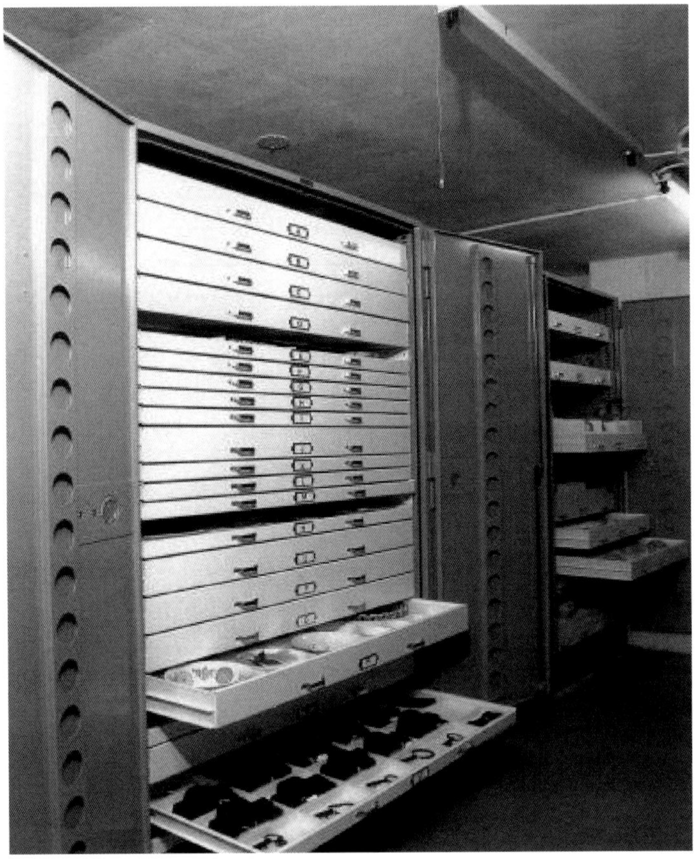

Figure 8
New art storage cabinets at the Seattle Asian Art Museum. The art is placed in custom-tailored foam trays. Photo: Paul Macapia.

shelves have straps attached with rivets. For ease of access, each strap has been furnished with a belt clip attachment in the center that can be unclipped when works of art need to be removed or viewed (Creahan and van Handel 1990). Large objects are properly padded with polyethylene foam and tied directly to the shelving unit or to a prepared wall panel with soft strapping.[25]

A recent development has been the incorporation of compact storage at SAAM and in the SAM building downtown. Although there are drawbacks to any storage system in which several units must be moved to obtain access to a single object, we made the decision to install compactor systems in several areas of storage as the growth of the SAM collections threatened to force space constraints upon us.[26] The possibility of exploring off-site storage was considered; however, this option was ultimately rejected by the SAM curatorial and conservation departments in favor of compact, in-house facilities. Close proximity to the collections was determined to be of paramount importance in an institution that already has multiple sites.[27]

The new compact storage system and paintings rack systems were designed and built to the latest code standards.[28] One significant effect of the change to the *International Building Code 2003* has been the tightening of guidelines relating to nonstructural elements. The first painting storage racks were installed in the downtown building in 1991; however, the system that had then been used to suspend the racks no longer met the more rigorous code requirements. Before new racks could be installed, the ceiling-mounted track for the hanging racks required upgrading with supplementary attachments to the ceiling. The newly installed painting racks have a horizontal bias to the grid and to the flat steel S hooks, which serve to reduce the amount of sway for stored paintings.

Although no measurements were made, observation of the conditions in storage during certain construction phases indicated highly effective vibration damping for objects on Ethafoam-lined shelves on the compactor system. Transfer of vibration to objects was greater for freestanding shelving units, and the most severe levels of transferred vibrations were noted, not surprisingly, on those units that were fastened to the wall.[29]

The Olympic Sculpture Park

The Seattle Art Museum's Olympic Sculpture Park, which opened in January 2007, represents a new direction for the museum, and here too, the design and construction teams were confronted with considerable seismic challenges (fig. 9).[30] These challenges included a seawall, a steep slope, and two large bridges; the nine-acre park has been developed on a sloping, regraded plot that is bisected by a major street and by the Burlington Northern Santa Fe Railway tracks. A pavilion building is perched at the highest point, in the southeast corner of the park. From here, a path zigzags down to a beach on the waterfront past important sculptures by Richard Serra, Mark di Suvero, Alexander Calder, Tony Smith, Louise Bourgeois, and others.

In order to provide a stable substrate for these sculptures, many of which are large, and to provide a safe environment for the public, steep slopes were constructed with benching to prevent sliding or erosion (fig. 10). The structural stability of these terraces was achieved by creating an underlying ground structure constituted of steel cages filled with rubble. The large, precast

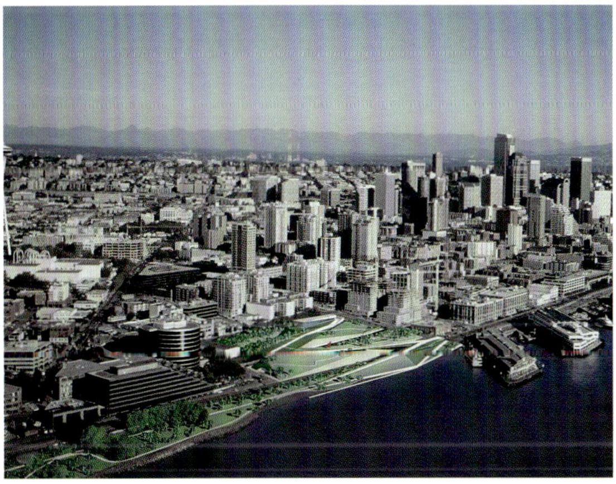

Figure 9
The Seattle Art Museum's Olympic Sculpture Park. A regraded waterfront site, traversed by major transit thoroughfares, provided challenges for the design architects and geotechnical engineers. Computer graphic element: Weiss/Manfredi Architects, courtesy of the Seattle Art Museum; aerial photo: Alan Abramowitz Studio.

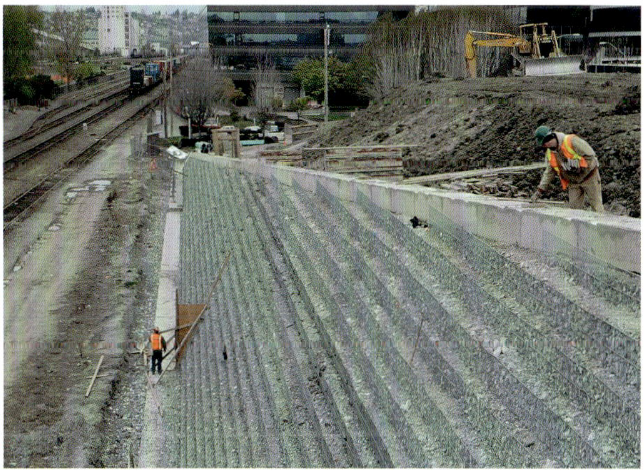

Figure 10
At Olympic Sculpture Park, a bulwark made of mechanically stabilized earth protects against subsidence of the park. Photo: Paul Macapia.

concrete panels that slope down to Elliott Avenue and the railway are anchored firmly in place from below to prevent movement.

A key area of consideration was the waterfront beach and seawall, because such areas (and the saturated, below-grade sand and gravel layers inshore from them) might be expected to be vulnerable to liquefaction and destabilization in an earthquake. The original seawall, constructed in 1934, was crumbling and in disrepair, and it was determined to be wholly inadequate for seismic protection, given the proposed construction. Considerable analysis of the local environment and of the design proposal was undertaken by geotechnical and aquatic engineers, so that provision could be made for adequate stability for this part of the park in a major earthquake. To ensure that the entire waterfront remain sufficiently secure, piles were sunk to support the seawall, and an embankment of rock at the shoreline was employed to compress and stabilize the ground, thereby arresting any potential slippage toward the sound of the abutment that supports the railway bridge.[31]

Each piece of art going into the park brings with it a number of conservation challenges. The conservation department at SAM has worked with professional art fabricators, art installers, and engineers to specify footing attachments that will ensure adequate stability for each sculpture, as well as specify an appropriate means of fastening the sculptures to the concrete footings.[32] Large sculptures such as Richard Serra's *Wake* must be secured in place for life safety and for preservation reasons; for such pieces, we have worked with the artists and with engineers to propose footing attachments that are effective and aesthetically acceptable. For smaller pieces, such as Tony Smith's *Wandering Rocks*, the need for tying down the component parts is determined as much by security issues as it is by seismic concerns.

Although a tsunami or seiche on Puget Sound might be expected to be relatively modest in scale, procedures are currently being developed for public safety on the waterfront precinct of the park, and permanent installations of sculpture have been limited to the areas of the park that lie east of the seawall until further information is known about potential risks.

Emergency Response

A crucial aspect in the negotiation of seismic threats is the preparation of appropriate emergency preparedness and response systems. The April 27, 2006, edition of the *Seattle Post-Intelligencer* newspaper carried the headline "There Is No Way We Are Ready." The story related to the publishing of a report from Blue Cascades III, a tabletop exercise, organized by the Pacific Northwest Economic Region. The exercise brought together representatives of 150 regional organizations to analyze preparedness for a Richter magnitude 9 megathrust earthquake on the Cascadia Subduction Zone and to assess the resilience of the regional infrastructure to such a quake. Findings included the fact that many attendees had inadequate emergency preparedness systems in place in their organizations and that many had little understanding of the physical effects that such a quake would have on their buildings. Even the experts present were unable to predict with precision the consequences of cascading and simultaneous regional destruction and the full extent of damage to the critical facilities of the region. Recommendations included seeking federal support for detailed research, accurate analysis of the impact of a sizable earthquake on the area's systems and interdependencies, and a need for organizations to establish procedures and to improve the resilience of critical infrastructure (King County Office of Emergency Management 2006).

These requirements—recognizing and studying the threat and implementing procedures and protective measures—also hold true if life and the collections in the museum are to be preserved in an emergency. SAM has an award-winning emergency response plan in which procedures for a number of emergency situations are described.[33] The museum also continues to be involved in training for emergency preparedness and response as the institution grows and our understanding of the seismic risks develops.[34] The report of the Blue Cascades III exercise indicates that adequate preparedness can only be achieved by dedicating considerable effort to comprehending the threats and their regional consequences. Both procedural and physical contingencies need to be implemented to mitigate those threats as far as possible.

Conclusion

A general museum such as the Seattle Art Museum faces a broad spectrum of seismic challenges. If our understanding of the range of potential seismic impacts has expanded, there remains a need for even more precise characterization of those threats and for thorough emergency planning. Preparedness must bridge the gap between our current, incomplete model of comprehension (based partly on recent quakes) and the potential occurrence of a devastating earthquake of great magnitude. We need effective emergency response systems, and as we enhance our museum structures, we must lessen potential damage to collections by further developing our construction standards so that both art and life will be protected. By assiduously maintaining mount making, storage, and display standards and by researching further the specific seismic forces to which art might be subjected, we will provide our collections with the optimum protection. We will also know that we have established the best achievable foundations for a speedy reinstatement of access to the museum following the inevitable massive earthquake.

Acknowledgments

The author wishes to thank Brian Atwater, Laura Ballock, John Bingham, Liz Brown, Julie Creahan, B. J. Farrar, Bob Freitag, Cascadia Regional Earthquake Network, Daryl Gardner, Jim Haarsager, Scott Hartley, John Hooper, Joy Jacobson, Sarah Kleiner, Gordon Lambert, Patricia Leavengood, Susan Lewandowski, Paul Macapia, Jack Mackey, Jerry Podany, Allison Rabbitt, Jay Taylor, the U.S. Geological Survey, and Heather Wirth.

Notes

1. The Seattle population was approximately 366,000 in 1933 (Kueter 2001), and it currently stands at approximately 572,600 (2004 estimated population from Seattle government website [City of Seattle 2005]). The Seattle Art Museum organization now administers the Seattle Asian Art Museum in Volunteer Park, the Seattle Art Museum in downtown Seattle, and the Olympic Sculpture Park on the Seattle waterfront. The collection currently stands at over twenty-two thousand works of art.

2. See Adams (1992) for an overview of scholarship on prehistoric seismic activity in the region.

3. For an account of the seismic profile of the Sumatra Earthquake, see Zimmerman et al. (2005, 8–9).

4. Ten thousand people were killed in the earthquake that severely damaged Mexico City, around 400 km (250 mi.) from the fault line. See Atwater et al. (2005, 8–9).

5. Bob Freitag, director of the Institute for Hazards Mitigation Planning and Research, University of Washington, conversation with the author, April 26, 2006.

6. Nearly a year after the hurricane, many thousands of people are still living in temporary accommodations far from their homes.

For an account of the regional damage caused by the hurricane, see Knabb, Rhome, and Brown (2005).

7. See Atwater et al. (2005).

8. Bob Freitag, director of the Institute for Hazards Mitigation Planning and Research, University of Washington, conversation with the author, April 26, 2006.

9. Bob Freitag, director of the Institute for Hazards Mitigation Planning and Research, University of Washington, conversation with the author, April 26, 2006.

10. For scenario studies, see Zimmerman et al. (2005, 8–9) and Earthquake Engineering Research Institute and the Washington Military Department Emergency Management Division (2005).

11. Bob Freitag, director of the Institute for Hazards Mitigation Planning and Research, University of Washington, conversation with the author, April 26, 2006.

12. For procedures in particular situations, see Seattle Art Museum (2001).

13. Jim Haarsager, maintenance supervisor, Seattle Asian Art Museum, conversation with the author, April 2006.

14. A shear wall provides sufficient strength and stiffness to resist horizontal forces in an earthquake. The shear wall anchors overlying structures and transfers the seismic loads to floors or foundations below.

15. The museum expansion was designed by Allied Works Architects (AWA) of Portland, Oregon. The architect of record for tenant improvements is LMN Architects of Seattle. The Washington Mutual Tower was designed by NBBJ of Seattle. The expanded museum opened in 2007 and will take over twelve floors of the AWA structure by 2020.

16. Jay Taylor and John Hooper, Magnusson Klemencic Associates, Seattle, conversation with the author, April 25, 2006.

17. The building has been constructed to preserve life safety during a Richter magnitude 9 earthquake on the Cascadia Subduction Zone and a magnitude 7.1 earthquake on the Seattle Fault.

18. Jay Taylor of Magnusson Klemencic Associates, Seattle, and Jim Boorstein of Traditional Line, New York, determined the most appropriate attachment hardware and the correct number of attachment points after the condition of the wood had been examined closely and the species had been identified, processes that yielded information on the density and mass of the panels.

19. The secure placement of this installation was determined by engineers at Magnusson Klemencic Associates, Seattle, and the installation team at SAM and Fabrication Specialties of Seattle.

20. Experimental rocking studies for objects and design guidelines, including designs for base isolators, are described in Agbabian, Masri, and Nigbor (1990).

21. Some material describing the approach of the SAM mount makers was presented in a booklet that was prepared for a one-day practical workshop, hosted by Jack Mackey, B. J. Farrar, and Gordon Lambert at SAM on June 27, 1997. See Mackey, Farrar, and Lambert (1997).

22. Jack Mackey, SAM mount making coordinator-preparator, conversation with the author, April 2006. See Mackey, Farrar, and Lambert (1997).

23. The isolator was designed and installed by B. J. Farrar of the J. Paul Getty Museum, in collaboration with SAM conservation staff and a crew from Artech, Seattle. The sculpture is a partial promised gift to the Seattle Art Museum collection.

24. The SAM conservation department is now composed of two full-time conservators, a coordinator, two collections care managers, an archival technician, and an intern.

25. Ethafoam—closed-cell polyethylene foam—and nylon strapping or unbleached cotton are generally used.

26. The high quality and smooth operation of the selected compactor systems were viewed and approved prior to purchase.

27. Funding for these projects was largely supported by the Seattle Art Museum capital budget. The museum recognized the need for safe, in-house storage that could accommodate anticipated growth in the collection. This augmentation of SAM storage capacity also enabled the museum to reduce the need to pack and move art and the need for temporary off-site storage during construction.

28. The new storage compactors and hanging racks at SAM and SAAM were designed and installed by Spacesavers Northwest of Seattle.

29. At certain stages in construction, neighboring structures were demolished, and holes were cut in the north wall of the SAM structure. Art was evacuated from storage that was adjacent to construction and placed in safer locations. Storage facilities were monitored continuously by SAM security and conservation staff. Accelerometers were employed to measure vibration levels in the initial stages of demolition, which took place while the museum was still open; measurements were taken to ensure that threshold levels were not exceeded for the art that remained on display.

30. Design architects for the Olympic Sculpture Park were Weiss/Manfredi Architects of New York.

31. John Bingham, Hart Crowser engineers, Seattle, conversation with the author, April 2006. The geotechnical consultants advised that the seismic conditions should be separated and that the embayment, the bridge abutment, and the general slope of the park should be considered separately. During a 2475 earthquake (that is, an earthquake that would be statistically likely during a 2475 year period), very little movement is expected at the Burlington Northern Santa Fe Railway bridge; minor movement is expected for the general slope of the park; and some movement is anticipated on the beach.

32. Appropriate footing attachments were determined by Magnusson Klemencic Associates, Seattle, and SAM conservation, with contributions from Fabrication Specialties of Seattle, Budco Enterprises of New York, and Lippincott's of North Haven, Connecticut.

33. Seattle Art Museum (2001); this emergency response handbook was complied by Barbara Roberts and the registrar of the Seattle Art Museum (the last update was in 2001). It is based on the "Emergency Planning Handbook" of the J. Paul Getty Museum and the recommendations of the Seattle Fire Department. Response procedures have also been updated to deal with specific potential threats relating to construction.

34. SAM hosted the American Institute for Conservation of Historic and Artistic Works (AIC) "Emergency Response for Cultural Institutions: Regional Emergency Response Workshop" in July 2001. The author was on the planning committee for the "Seattle Alliance for Response Forum" at Seattle's Museum of History and Industry in 2006. SAM mount makers and conservation staff attended "Protecting Collections from Earthquake Damage" at the J. Paul Getty Museum in May 2006, and SAM conservation plans to host another AIC meeting on emergency response in 2007.

References

Adams, J.
1992 "Paleoseismology: A Search for Ancient Earthquakes in Puget Sound," *Science* 258 (Dec.), pp. 1592–93.

Agbabian, M. S., S. F. Masri, and R. L. Nigbor
1990 *Evaluation of Seismic Mitigation Measures for Art Objects* (Los Angeles).

Atwater, Brian, et al.
1995 "Summary of Coastal Geologic Evidence for Past Great Earthquakes at the Cascadia Subduction Zone," *Earthquake Spectra* 11, no. 1, pp. 1–18.

Atwater, Brian, Musumi-Rokkaku Satoko, Satake Kenji, Tsuji Yoshinobu, Ueda Kazue, and David K. Yamaguchi
2005 *The Orphan Tsunami of 1700: Japanese Clues to a Parent Earthquake in North America* (Seattle). City of Seattle
2005 *Seattle-at-a-Glance* (Seattle's official census site), http://www.seattle.gov/dpd/Research/Population_Demographics/Overview/default.asp (accessed April 9, 2007).

Creahan, J., and L. van Handel
1990 "Seismic Protection in Museum Storage: Affordable Solutions," *WAAC* [Western Association for Art Conservation] *Newsletter* 12, no. 3 (Sept.), pp. 23–25, http://palimpsest.stanford.edu/waac/wn/wn12/wn12-3/wn12-310.html.

Earthquake Engineering Research Institute and the Washington Military Department Emergency Management Division
2005 *Scenario for a Magnitude 6.7 Earthquake on the Seattle Fault*, ed. Mark Stewart, http://seattlescenario.eeri.org/documents.php (accessed April 9, 2007).

Fuller, Richard E.
1993 *A Gift to the City: A History of the Seattle Art Museum* (Seattle).

King County Office of Emergency Management
2006 *Infrastructure Interdependencies Tabletop Exercise, Blue Cascades III: Managing Extreme Disasters* (Bellevue, Wash.), http://pnwer.org/pris/finalrep.pdf.

Knabb, Richard D., Jamie R. Rhome, and Daniel P. Brown
2006 *Tropical Cyclone Report: Hurricane Katrina: 23–30 August 2005*, www.nhc.noaa.gov/pdf/TCR-AL122005_Katrina.pdf (accessed June 22, 2007).

Kueter, Vince
2001 "Seattle through the Years," *Seattle Times*, Nov. 13, http://seattletimes.nwsource.com/news/local/seattle_history/articles/timeline.html (accessed April 9, 2007).

Leavengood Architects
1995 "Occupant and Exit Analysis and Seismic Evaluation for Seattle Asian Art Museum," Seattle Art Museum.

Leslie E. Robertson Associates
1989 "Study of Seismic Response of Plinth for Two Guardian Sculptures," Seattle Art Museum.

Mackey, Jack, B. J. Farrar, and Gordon Lambert
1997 "Mountmaking Workshop at Seattle Art Museum," Seattle Art Museum.

Podany, J.
1989 "Earthquake Research Applied to the Installation of Two Monumental Sculptures," Seattle Art Museum.

Seattle Art Museum
2001 "Seattle Art Museum Emergency Response Handbook," Seattle Art Museum.

Zimmerman, Robert, Gail Dreckman, Timothy Walsh, Stacy Bartoletti, Bob Freitag et al.
2005 *Cascadia Subduction Zone Earthquakes: A Magnitude 9.0 Scenario*, http://www.crew.org/papers/CREWCascadiaFinal.pdf (accessed April 19, 2007).

Case Studies: The Conservator's Role in Seismic Mitigation of Artwork at the Los Angeles County Museum of Art

John W. Hirx

Introduction

Protecting works of art from damage during a seismic event at the Los Angeles County Museum of Art (LACMA) is an ongoing, ever-evolving activity. Mounts created to protect LACMA's collection during earthquakes tend to fall into two general categories: mounts that can be made and installed by the art preparation staff or mount makers, and mounts that are designed by conservators or engineers, built by fabricators, and installed by conservators.

Those works of art that do not require alteration for installation on a mount are the responsibility of the Art Preparation and Installation staff. Small works of art with low mass are not considered at high risk in a seismic event. Microcrystalline wax is commonly used to hold lightweight objects in place on exhibition decks. The wax can be applied and removed from any nonporous or vitreous surface such as ceramic or glass. Though wax has been used by conservators for years in this capacity, recent lab studies have confirmed the use of microcrystalline wax in this context over the use of other waxes (see the paper "Performance Expectations for Microcrystalline Waxes for the Seismic Protection of Art Objects" by Crowley and Laefer, included in the present volume). Wax and padded brass mounts, such as pins, clips, brackets, or T-mounts with monofilament to tie and secure the object, account for the greatest number of mounts used to support and seismically stabilize works of art at LACMA, all of which tend to be small to medium sized. These low-tech approaches do not require long time periods for concept development, costly engineering, or alteration of the object.

Large objects, often life-size or larger, such as stone sculptures, are high mass and high risk, and they frequently require invasive seismic mitigation techniques, which usually include drilling and installation of hardware in order to attach the object to a mount or base isolator.

Custom-Fit Mounts and the Introduction of the Base Isolator

Prior to the Northridge Earthquake of January 14, 1994, large objects were exhibited in LACMA's galleries with conventional mounts, which are usually external and visible. In addition, other mitigation techniques for large sculpture included the use of ballast in the bottom of display bases to lower the center of gravity of the object-base system and the practice of placing the statue and its base not directly on the floor but on a piece of Lennite, an ultra-high-density polymer with a smooth and slick surface. The product allows horizontal movement of the sculpture should an earthquake occur, and it reduces the risk of damage. These practices continue to this day.

Base isolators had yet to be used at LACMA because of the high costs associated with their engineered design, manufacture, and installation. One low-cost intermediate mount strategy that can be implemented until funding becomes available to purchase a base isolator is to mount objects as low as possible on weighted bases for which the width and depth are maximized, thereby lowering the center of gravity of the entire system and controlling, as much as possible, its overall overturning moment during a seismic event. This system is still in use today, modified as needed for practical purposes. All large-scale sculptures were installed on hollow plywood bases, and mounting hardware already on the object at the time of acquisition was often reused. Lead and iron bars and lead pellets were installed in the wooden bases as ballast to lower the center of gravity of the overall object-mount-base system.

In the early to mid-1980s, in order to determine relative base dimensions and ballast needs, the BASTAT formula was developed. The BASTAT formula, which derives its condensed name from *ba*llast and *stat*uary, is a simple mathematical equation used to estimate the amount of ballast (lead bricks, etc.) needed to lower the center of gravity in the system of an object and affixed base to withstand a certain standard horizontal force before toppling. When the formula is used, the object and its attached base, which covers a ballast box containing ballast, are considered as a monolithic system. For information on similar formulas, refer to the Getty publication *Evaluation of Seismic Mitigation Measures for Art Objects* (Agbabian, Masri, and Nigbor 1990) (see appendix 1).

Lennite, a trade name for ultra-high-molecular-weight polyethylene (UHMW), used in sheet form, is attached to the underside of the ballast box. The material, somewhat similar in character to Teflon or Delrin, is slippery. During a seismic event, if necessary, the object and its base may slide slightly, partially isolating the artwork from ground motions. In addition, Lennite has been extensively used throughout LACMA as a low-cost isolation solution for freestanding exhibition cases that are not bolted to the floor. All of these cases have ballast boxes filled with weights, and a 6.3 mm (¼ in.) thick sheet of Lennite is attached to the underside of the box bottoms; it allows the cases to move during an earthquake.

As a result of past mitigation measures, only a few objects in LACMA's collection sustained significant damage during the Northridge Earthquake. However, the damage incurred justified funding a program to purchase base isolators engineered and fabricated for designated objects in the collection. The three individuals responsible for the implementation of LACMA's base isolation program are Steve Colton, former head objects conservator, LACMA; Pieter Meyers, former head of conservation, LACMA; and Jack Yaghoubian, an engineer and the inventor of the base isolator for LACMA's artwork.

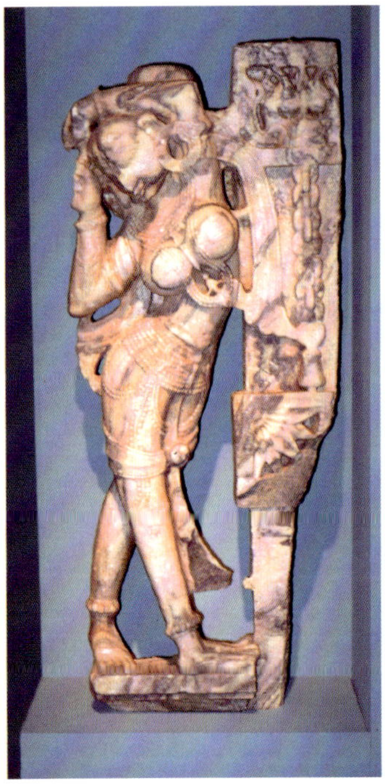

Figure 1
Celestial Nymph
India, Rajasthan, ca. 1450
Marble, 112.4 × 50.8 × 24.1 cm
(44¼ × 20 × 9½ in.)
Los Angeles County Museum of Art
M.80.62
Gift of the Ahmanson Foundation
Photo: © 2006 Museum Associates/ Los Angeles County Museum of Art.

Case Study One: *Celestial Nymph*

Conservators become involved in the mount making or mount design process when an object is very large and/or heavy and must be altered in some manner. LACMA's *Celestial Nymph* (fig. 1), a 113.4 kg (250 lb.) marble sculpture, is a female figure who stands on a ledge attached to a post. The post terminates below the sculpture as a tenon. The former mount consisted of a block of wood, wide enough to support the length and depth of the sculpture, with a hand-carved hole to accommodate the tenon. A wood plank extended upward from the support block, up the back of the post, from which a small clip extended to support the upper half of the object. In order to build a new, more substantial mount, a template had to be created. The template made it possible to fabricate a hollow form-fit stainless steel mount.

The new mount supports the weight of the piece and accommodates the lower tenon. The hollow box at the bottom of the mount has permanently installed bolts extending from the bottom which allow it to be mounted to timber framework in a gallery. The hollow cavity of the box has been filled with foamed epoxy, Ciba-Geigy epoxy GY 6010 (resin)/HY956 (hardener) foamed with 0.02% polymethylsiloxane, which surrounds the tang, preventing lateral movement. PhillySeal R sealing and fairing epoxy putty was adhered to the mount to serve as a spacer between the mount and the nonuniform upper surface of the back of the piece.

Three threaded inserts were adhered into the back of the sculpture. Bolts pass through recessed holes in the mount and connect to the threaded inserts, thereby mechanically connecting the object to the mount so that the object and mount function as a single unit that can be exhibited, sent on loan, or stored without disassembly. The mechanical nature of the assembly makes disassembly of the object from the mount quite straightforward. The mount is not apparent in figure 1 because the object hides part of it and the rest is painted to match adjacent walls.

Years of direct observation by staff have revealed that the top floor of the Ahmanson Building moves the most during earthquakes. Today all isolators used at LACMA, which can be installed in any gallery, are engineered for the parameters of the top-floor galleries, as the worst-case scenario.

The ten base isolator systems in use at LACMA each consist of a self-centering plate and roller system, for a total of nine moving parts. Between the bottom and middle plates, three parallel tracks are mounted. Between the middle and top plate, three more tracks are mounted perpendicular to those below the middle plate. Side rails prevent the plates from fully separating. Figure 2 shows an isolator fully open and held in position by Colton and Yaghoubian.

The two sets of perpendicular tracks allow motion in the two horizontal planes (x and y). As the rollers move along the tracks, they are held in position as a group by pairs of sidebars. The tracks are not flat; they have shallow valleys in which the rollers can slowly roll up and down. At the end of a seismic event, the rollers return to their resting position at the bottom of the valleys. This feature constitutes the centering mechanism.

When the earth moves, the plates and rollers in the isolator move or displace to isolate the artwork from seismic motion, thereby preventing it from falling during an earthquake. The plates typically displace up to 40.6 cm (16 in.) side to side with a period of 4 sec. The displacement parameter is derived from data that describe the seismicity of the environment—information found in soil reports, oil exploration documents, and drillings performed when the LACMA buildings were constructed. All isolators move on the x and y horizontal axes; there is no isolation of the z axis, the vertical direction. When the isolator is at rest, there is no force on either the x or y horizontal axis.

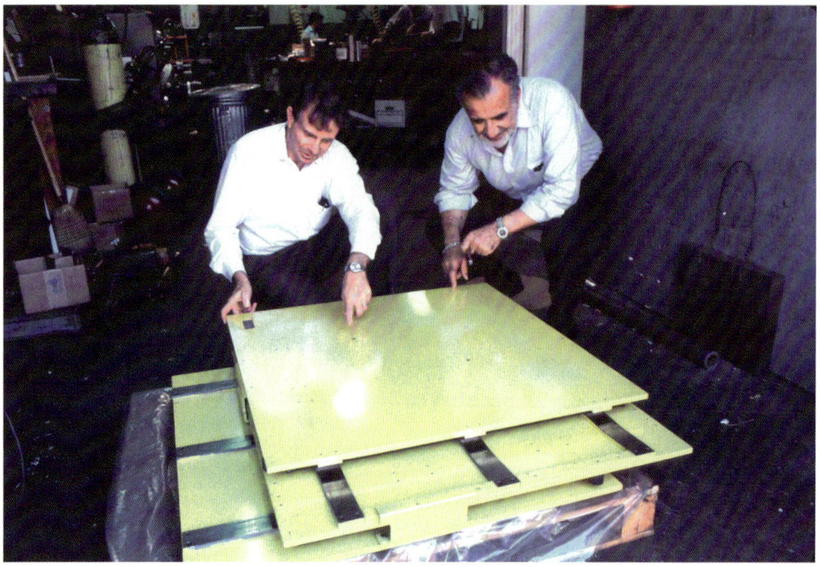

Figure 2
Base isolator fully opened, with the perpendicular tracks exposed. The isolator is held in position by Steve Colton (left) and Jack Yaghoubian (right). Photo: © 2006 Museum Associates/ Los Angeles County Museum of Art.

During a seismic event, ground motion causes the plates to displace in a side-to-side fashion. However, if forces were strong enough to cause an entire object-isolation mechanism to move up and down along the z axis to an extent far exceeding 1.3 g, it is our opinion that such forces would probably destroy the building and its contents. Everything is already withstanding a 1 g vertical load in the steady state, and most objects can easily sustain an additional 0.3 g vertical load, for a total of 1.3 g, without any significant trauma. (After all, this 30% greater load is less than that experienced in a normal elevator ride.) Therefore, z axis isolation is not a component of this system.

Although the ten isolators currently used by LACMA are of the same type of mechanism, they vary in the materials from which they are made (such as steel or aluminum), in overall size, and in weight. They have been engineered as one or two units. An isolator can be made inclusive or exclusive of an intermediate mounting plate, which may also serve as a counterweight. Engineering an isolator in this way can be beneficial for shipping an object to an area that is not seismically active and therefore does not need the full isolation system for display.

Case Study Two: *Nature Goddess*

LACMA's *Nature Goddess* (fig. 3) is a larger-than-life-size sculpture carved from a single block of Mathura red sandstone. Previously the object was mounted for exhibition with steel rods extending down from both legs into the pedestal and a waist-style mount that attached to the wall. As a typical example of LACMA's seismic mitigation program, the sculpture was transferred from its old mounts to a base isolator, and it is currently on display in the gallery with new stainless steel rods connecting the sculpture to the isolator. Figure 3 shows the sculpture moving from side to side on its base isolator.

The replacement rods are of larger diameter and made of a stronger alloy than the previous rods. The old rods were removed with a water-cooled core diamond tipped drill bit. The core drill bit had a larger inside diameter than the rods. It was chosen in order to enlarge the diameter of the holes, as well as

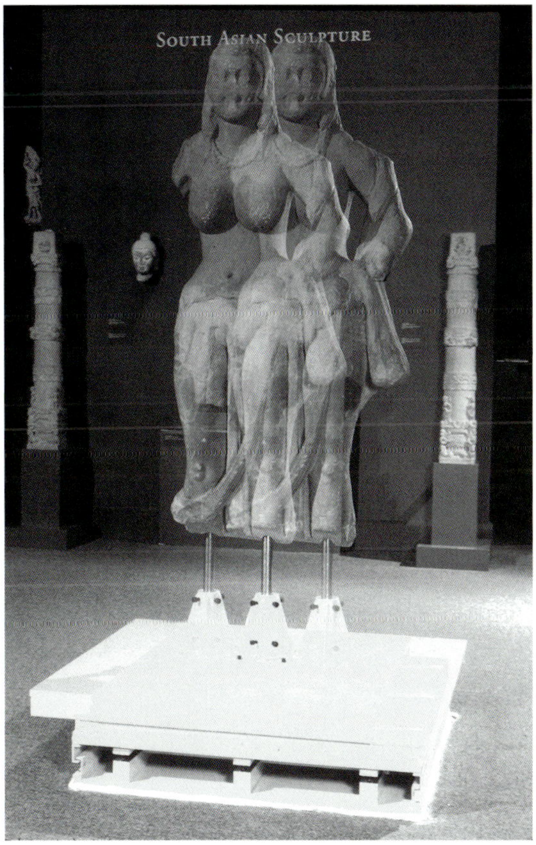

Figure 3
Nature Goddess
India, Uttar Pradesh, Mathura,
1st century
Mottled red sandstone, 160.0 × 58.4 × 31.7 cm (63 × 23 × 12½ in.)
Los Angeles County Museum of Art
M.86.21
Purchased with funds provided by Mr. and Mrs. Allan C. Balch
Photo: © 2006 Museum Associates/Los Angeles County Museum of Art.

extend the holes' depth into the thighs of the sculpture to accommodate larger-diameter and longer stainless steel rods to provide a stronger mounting system. Prior to adhering the rods in place with Ciba-Geigy epoxy GY 6010 (resin)/HY956 (hardener) foamed with 0.02% polymethylsiloxane, multiple applications of Rohm and Haas Acryloid B-72, an ethylmethacrylate-methylacrylate copolymer in acetone, were applied inside the newly drilled holes. The rods, and therefore the sculpture, are secured in a mount on the top of the isolator. The rods are visible; the mount and isolator are covered. For maintenance purposes, the LACMA isolators need to be slightly pushed periodically to help prevent point loading on the rollers and ensure their freedom of movement.

Reused mount hardware has often proved inadequate. Over time, many sculptures, such as *Nature Goddess*, have been repeatedly moved. After the Northridge Earthquake, supplemental or remedial mounts were installed on objects in jeopardy, including *Nature Goddess*. In this case, a harness-style mount was put in place around the sculpture's waist to prevent it from falling forward in a seismic event. At a later point the base isolator was installed.

Case Study Three: Viola Frey's *Sky Blue Suit*

In 1997 LACMA acquired a large-scale, low-fire ceramic sculpture entitled *Sky Blue Suit* (fig. 4) by Viola Frey. Frey was an Oakland, California, sculptor who in the 1980s moved away from making solid tabletop-size sculptures. Today she is known for her hallmark over-life-size, hand-built, hollow, multisectional figures that are colorfully glazed.

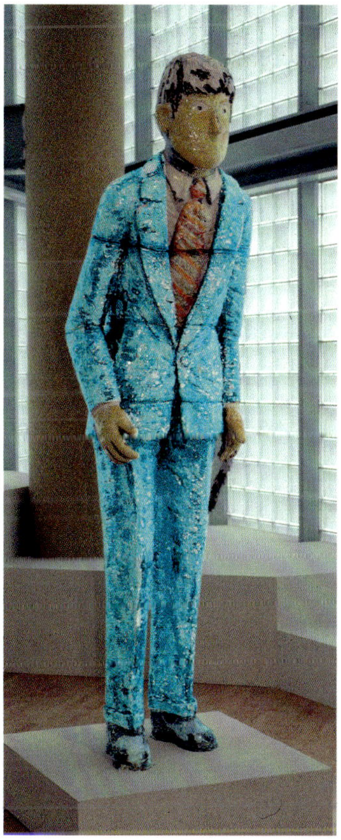

Figure 4
Viola Frey
United States
Sky Blue Suit, 1982
Ceramic, 250.2 × 68.6 × 55.9 cm (98½ × 27 × 22 in.)
Los Angeles County Museum of Art
AC1997.117.1.1-.6
Smits Ceramic Purchase Fund
Photo: © 2006 Museum Associates/Los Angeles County Museum of Art.

All of Frey's oversize sculptures are similarly made. Figures and objects are divided into separable sections. In the case of *Sky Blue Suit*, six stacked and open-ended sections make up the piece. The head section is closed at the top, and the soles of the shoes are open ended, so that access to the entire interior of the assembled sculpture is provided. The finger marks from hand building are not only visible, they are a distinct texture, as are the drips and brushstrokes from the glazing.

Sky Blue Suit is approximately 2.5 m (over 8 ft.) tall; it is top-heavy and leans forward. The last characteristic contributed to the sculpture's candidacy for a base isolator. Figure 4 shows the sculpture displayed at LACMA on its covered base isolator during the 2000 exhibition *Color and Fire: Defining Moments in Studio Ceramics, 1950–2000* (Lauria et al. 2000).

Deep unglazed horizontal ledges extend inward approximately 5 cm (2 in.) at the top and bottom of the vertical walls of each section, which have holes that are used to secure the sections together with nuts and bolts (fig. 5). The ledges were also used during manufacture of the sculpture as surfaces upon which to stack the sections in the kiln. The colorful glazes blended at the time of application and during firing in the kiln. During firing, the glazes became fluid enough to flow and gather as drips at the junction of the bottom of the walls and external edges of the ledges, indicating that the pieces were elevated in the kiln.

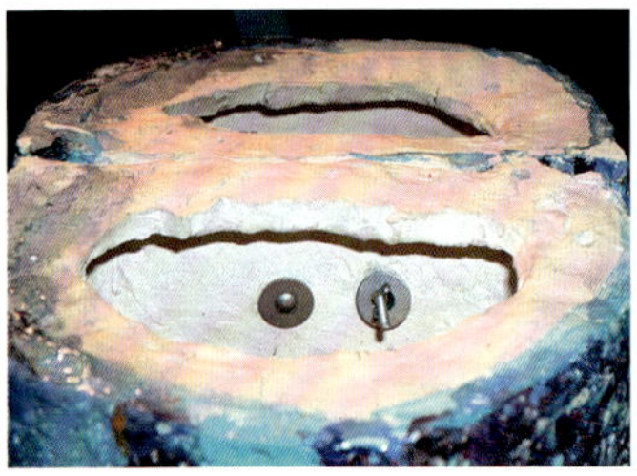

Figure 5
Original nuts and bolts used to hold leg sections of *Sky Blue Suit* together. The unglazed ledges are shown. Photo: © 2006 Museum Associates/ Los Angeles County Museum of Art.

Frey often used white silicone to attach her multiple-unit figures together. In the case of *Sky Blue Suit*, silicone was used to adhere the head section to the neck socket; it was not used elsewhere in the sculpture. When the sculpture was acquired by LACMA, mismatched nuts and bolts had been used by Frey to align holes of the abutting ledges to connect the sections together, and small sheets of Plexiglas were placed in between sections as spacers. This assembly technique presented a number of drawbacks. Over the years, the glaze drips between the stacked sections at the wall-ledge junction had abraded against one another, causing the glaze to spall and chip.

Sky Blue Suit was originally mounted in a simple manner. It is likely that when the sections were initially assembled, after firing, the glaze drips around the soles of the shoes caused the sculpture to stand unevenly. Thick epoxy putty was applied to the soles of the shoes and served to level the figure's stance. Steel leveling plates were also adhered to the bottom of the shoes with the same epoxy. Two holes that extend laterally from the sides of the toes and heels of each shoe serve as points of attachment. Coarse threaded screws were used to secure the steel shoe plates to a wood plinth.

Once the sculpture was acquired by LACMA, it was disassembled at the previous owner's home and taken to the museum. The first step in the dismantling process involved cutting through the white silicone with a blade in order to release the head section. After the head was detached, it was easy to see that silicone remained bonded to the ceramic. Even though silicone was the adhesive of choice for the artist, because it is not easily reversed, it was not considered prudent to reemploy that method. After the head was removed, each section was disconnected from the section below by removal of the internal hardware by which the sections were held in position. During deinstallation, a large gap that had developed between the steel shoe plate and epoxy putty was noted. The gap indicated that the steel was no longer bonded to the epoxy. This was yet another reason to consider a base isolator for the support and protection of the object.

Sky Blue Suit was displayed at LACMA in the former Anderson Building, which has an outdoor gallery overlooking the main entrance corridor. The outdoor gallery has been primarily used for the display of ceramics of architectural scale. *Sky Blue Suit* was installed in much the same manner as it was formerly installed in a private collection. The shoes were screwed to the top of a low wooden buildup with four screws, using the four holes in the artist's steel shoe plate. The sections were again secured together with mismatched internal hardware. Nothing prevented abrading and chipping of the brittle glaze during an earthquake. When the sculpture was gently pushed, one could easily see the piece flex as each section rocked against the other, and the sound of glaze and ceramic cracking and grinding was equally clear.

Sky Blue Suit satisfied many criteria that made it a candidate for installation on a base isolator, not least of which was that it was a 2.5 m (over 8 ft.) tall, 204 kg (450 lb.), brittle, low-fired ceramic. The movement of each section against the other had great potential for damaging the ceramic. *Sky Blue Suit* had been included on the checklist for the upcoming traveling exhibition *Color and Fire*. It was merely a matter of time before it suffered more handling damage or suffered greater damage during a seismic event. These factors, in conjunction with the possibility of complete loss of the artwork in a significant earthquake, clearly called for the use of a base isolator.

Jack Yaghoubian was contracted to engineer the isolator for *Sky Blue Suit*. While the isolator was being built, three independent yet connected devices had to be developed in the lab: bumpers to cushion the assembled sections, an internal armature to join the sections together, and a method by which to attach the sculpture to the isolator. In short, *Sky Blue Suit* needed an interconnecting internal armature that would link each section, thereby creating a monolithic mass rather than a series of segments that would behave independently in a seismic event.

Creating the Internal Armature for *Sky Blue Suit*

The interior walls of each section of the sculpture were lined with fiberglass cloth adhered in place with concentrated solutions of Acryloid B-72 in various organic solvents, such as xylene, toluene, acetone, and ethanol. For the purposes of adhering the fiberglass in place, large quantities of B-72 were needed; however, specific resin concentrations and solvent choices were not germane to this aspect of the treatment. The lining made the inside smooth and served as a barrier layer to prevent the foamed epoxy (serving as a spacer between the interior wall of the sculpture and Ethafoam blocks) from migrating into the low-fire ceramic.

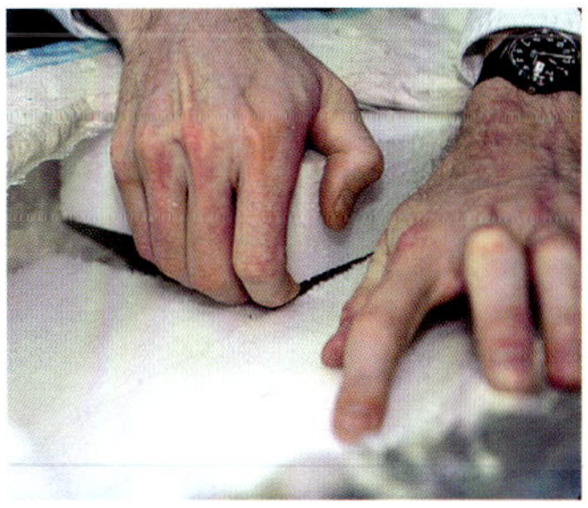

Figure 6
Installation of Ethafoam blocks. The internal cavities of each section of *Sky Blue Suit* had first been covered with a fiberglass cloth adhered in position with Acryloid B-72. Gaps between the blocks and the walls of the cavity were later filled with a foamed epoxy in order to consolidate the mass. Photo: © 2006 Museum Associates/ Los Angeles County Museum of Art.

After the fiberglass was installed, blocks of Dow Corning high-density Ethafoam HS-900 were cut and fitted to fill the hollow sections (fig. 6). Gaps between the Ethafoam and fiberglass were filled and consolidated with Ciba-Geigy foamed epoxy. Removable Ethafoam plugs, installed by pressure without the use of an adhesive, were created and inserted at the top of the leg sections, to facilitate assembly and disassembly of the sculpture.

Soft conforming silicone bumper-spacer inserts were created and installed between each section, replacing the hard, artist-installed Plexiglas sheet spacers, thereby preventing the remaining glaze drips from chipping and spalling. The bumper-spacer inserts consist of a sandwich of 1 mm (0.039 in.) thick Darteck (Nylon-66) and Dow Corning Neutral Cure Silicone 737 caulking.

A bottom sheet of 1 mm (0.039 in.) thick Mylar was laid on the top of a section of the sculpture to be protected. Copious amounts of silicone caulking were extruded onto the Mylar, covering the unglazed ledge. Wooden swab sticks were placed in the silicone around the outer perimeter of the ledge where the gaps between sections were narrowest, thus establishing a minimum spacing between abutted sections (fig. 7). A top sheet of Mylar was placed on the silicone. The upper section was lifted into place, and the excess silicone squeezed outward. It was then left to cure fully, and the wooden sticks were removed.

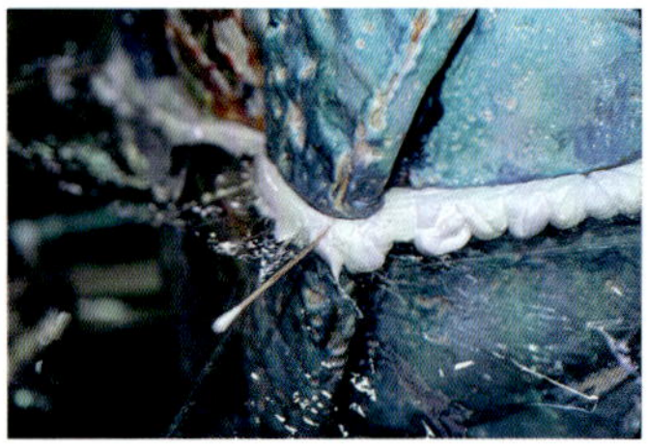

Figure 7
A swab, placed for spacing, embedded in the excess silicone of the bumper-spacer insert between two sections of the sculpture. Photo: © 2006 Museum Associates/ Los Angeles County Museum of Art.

After the silicone caulking cured, the inserts were trimmed and painted with undiluted Golden acrylic emulsion paints. The bumper-spacer inserts are not adhered in position. Because they were cast and pressed in place on the ceramic, they conform to the irregular surfaces. As a result, they all fit snugly in place. The silicone inserts serve to protect the abutted glazed edges during installation and deinstallation.

The sequence of the creation of the bumper-spacer inserts is as follows. The first silicone insert was created and installed between the legs. The second insert, installed on top of the legs and under the lower arms/hands section, had to be created at the same time as the first insert in order to maintain registration between the sections. After the first and second bumper-spacer inserts were created and the legs were lined with fiberglass, Ethafoam HS-900 plugs were fitted to the top inside of each leg. Short threaded stainless steel rods, approximately 7.6 cm (3 in.) each, were adhered in the soles of the shoes in the toe area. The two threaded rods extended through two holes in the aluminum mounting plate, and were secured in place with countersunk nuts, washers, and isoprene gaskets. The Ethafoam plugs were put in place inside the legs. Threaded stainless steel rods measuring 91.4 cm (3 ft.) in length were inserted; they passed through each leg and heel and the mounting plate, securing the leg sections to the plate. Countersunk stainless steel nuts and washers and isoprene gaskets were installed to secure the rod on the underside of the plate, with opposing nuts and washers compressing the HS 900. This method of attachment made one leg strong enough to stand on the mounting plate. The remaining sections of the figure were stacked on the mounted legs, connecting to the upward extending stainless steel rod shown in figure 8.

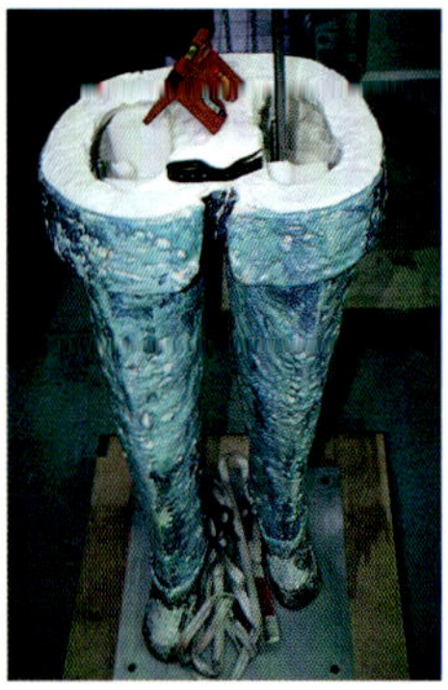

Figure 8
The legs of *Sky Blue Suit* bolted to the mounting plate. The stainless steel rod extends upward for connection to the next section. Photo: © 2006 Museum Associates/ Los Angeles County Museum of Art.

The use of eleven threaded stainless steel rods of varying lengths made it possible to connect the sections in order to create an interconnecting armature. In figure 9, superimposed upon the sculpture is a series of lines that illustrate how the threaded rods connect the sections. In practice, the rods run parallel to one another. Holes drilled through the Ethafoam accommodate the threaded rods. Nuts and washers secure the extreme ends of the rods onto the Ethafoam, securing two sections of the sculpture together at a time. The head has a 61 cm (2 ft.) long rod that extends from the Ethafoam and inserts into a hole inside the neck cavity. It is not secured with a nut and washer.

After the interconnecting armatures and bumpers were made, the sculpture was assembled on the isolator in the Objects Conservation lab at LACMA to fine-tune each section's fit. The leg and aluminum mounting plate assembly was initially bolted to the top of the isolator. All connections were hand tightened. This level of tightness was determined by the movement of the sculpture on the isolator. If a section rocked on the moving isolator, the nuts on the armature were tightened until, eventually, the section ceased to rock. This procedure was followed from bottom to top.

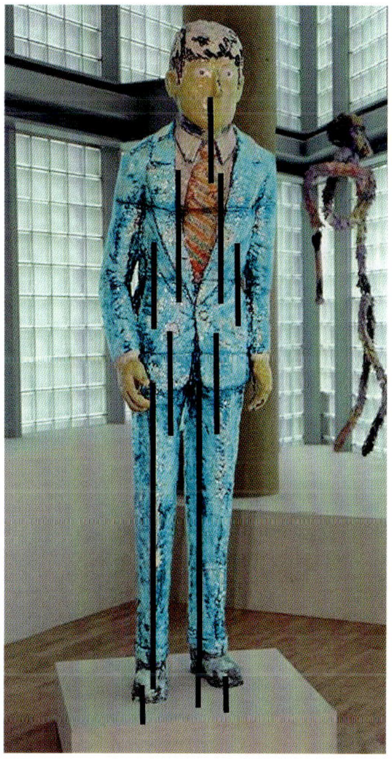

Figure 9
The assembled sculpture, *Sky Blue Suit*, with superimposed black lines that illustrate how the threaded rods connect the sculpture's sections. Photo: © 2006 Museum Associates/ Los Angeles County Museum of Art.

Installing the Object and Isolator in the Gallery Isolators are usually very large and can be extremely heavy, weighing upwards of a ton or more, depending upon the sculpture that is to be installed. A forklift is typically required to lift an isolator into position, after which point it is leveled. If the isolator is not level, it will not displace and return to center properly in an earthquake. All of LACMA's isolators are installed on a thin bed of level sand. Once the isolator is placed on the sand, the space between the underside of the edge of the lowest plate of the isolator and the floor is caulked with a bead of silicone to prevent the sand from migrating outward. During a seismic event, if the isolator needs to extend more than its full displacement, it can tear the silicone and slide on the sand to displace even further. In this situation, the sand functions as a pseudo-plate.

As in the lab, after the isolator is set, the legs with silicone insert are attached to the aluminum mounting plate with the previously described hardware system. The plate is then lifted into position on the isolator, and the upper sections are added and secured in place.

Isolator Cover The isolators installed in LACMA's galleries are covered and are therefore not apparent to the museum visitor. Covers on isolators can vary greatly in appearance based on gallery design, space around the sculpture, aesthetic issues, and so on. Functionally, the cover can be built large enough to allow full displacement of the isolator within the cover, or it can be built to conform closely to the outside dimensions of the isolator with movable sides that separate from the isolator during a seismic event, in an assembly called a breakaway system. For *Nature Goddess*, the four breakaway sides are L shaped, with ultra-high-density polyethylene sheeting attached to the bottom of the sides. During a seismic event, the sides will easily slide across the carpet, allowing the isolator to displace as necessary.

The sides of the cover for the isolator of *Sky Blue Suit* can be imagined as suspended upside-down drawers. A plywood surround is attached to the top of the aluminum mounting plate. The plywood serves as the attachment point for drawer slides. Each side of the cover is mounted on the drawer slides, which are bolted to the plywood. In an earthquake, as the isolator moves, the sides will push open like drawers, allowing the isolator to freely displace. After the event, the sides can be pushed back into place. For the exhibition *Color and Fire*, the cover was fitted very closely to the isolator to maximize the space around the sculpture.

The Future

Seismic mitigation at LACMA is an ongoing, ever-evolving activity. It is hoped that at some point in the future, LACMA will be able to undertake a full geological study to understand and update knowledge about the seismology of its property and surrounding land and faults (a study similar to one done for the J. Paul Getty Museum—Lindvall, Richter and Associates 1984). Such a study would greatly assist in improving the engineering of the base isolators used by LACMA. It is worth noting that the isolators have been doing their jobs. *Nature Goddess* displaced enough during a small recent earthquake to create gaps at the corners of the isolator cover, and yet the sculpture remained intact.

References

Agbabian, M. S., S. F. Masri, and R. L. Nigbor
1990 *Evaluation of Seismic Mitigation Measures for Art Objects* (Los Angeles).

Lauria, Jo, et al.
2000 *Color and Fire: Defining Moments in Studio Ceramics, 1950–2000* (Los Angeles).

Lindvall, Richter and Associates
1984 "Geologic and Seismic Safety Evaluation of the Main Museum Building for the J. Paul Getty Museum, Malibu, California," Department of Antiquities Conservation, J. Paul Getty Museum, Los Angeles.

Weidner, Richard, and Robert Sells
1965 *Elementary Classical Physics*, vol. 1. (Boston), p. 330, example 12.

Young, Hugh, and Roger Freedman
2004 *Sears and Zemansky's University Physics: With Modern Physics*, 11th ed. (San Francisco).

Appendix I

When an object is weighty and top-heavy, simple external mounts may not provide enough protection to prevent it from overturning during an earthquake. One method that can be used to develop and design safe exhibition parameters for a rigid system of object and base, or even for a freestanding vitrine containing objects, uses the BASTAT equation (a formula used to calculate the *ba*llast needed to stabilize *stat*uary):

$$B = \frac{DF_O(Y_1) + DF_B(Y_3) - RF(X)}{X}$$

The BASTAT equation is derived from the general equation

$$\Sigma MT = 0$$

The general equation states that the total sum (Σ) of the moments of inertia, defined as M = force × distance, about a point T, for a system (object and base) must be equal to 0 to be stable. If the inertia of the system is overcome and the rigid system is permitted to torque or rotate about point T (see fig. A.1), the potential for overturning is high. Therefore, the BASTAT formula can be used to model design parameters that decrease the system's ability to rock and torque to the point of instability in a given seismic event.

By applying known and desired numbers into the BASTAT formula, such as calculated toppling forces (DF_O and DF_B, the driving forces to topple an object and base), resisting force (RF), the center of gravity of the installed object and base heights ($Y1$ and $Y3$), and a desired base dimension (X), the formula can then be used to calculate ballast needs (B). Often design parameters, which call for a base with a small footprint, require impossibly large amounts of ballast. Therefore, several calculations need to be performed, with changes in the height at which an object is displayed, in the size of the base to which it is attached, in the minimum amount of ballast that should be placed in the base to lower the center of gravity of the system to counteract overturning, and in the level of gravitational risk in terms of the amount of g force the object is likely to be subjected to during an earthquake.

To use the equation for statuary or sculpture, certain assumptions must be made:

1. The object and base are rigid and rigidly attached to each other, so that they behave as a unit.

2. The object to be stabilized is *not* exceedingly top-heavy; thus:

$$Y_1 = \frac{\text{object height}}{2} + Y_2$$

3. The object is roughly centered on its base; thus: X = one-half least base dimensions on the floor.

4. The ballast does not significantly alter the center of gravity of the base; thus: $Y_2/2 = Y_3$.

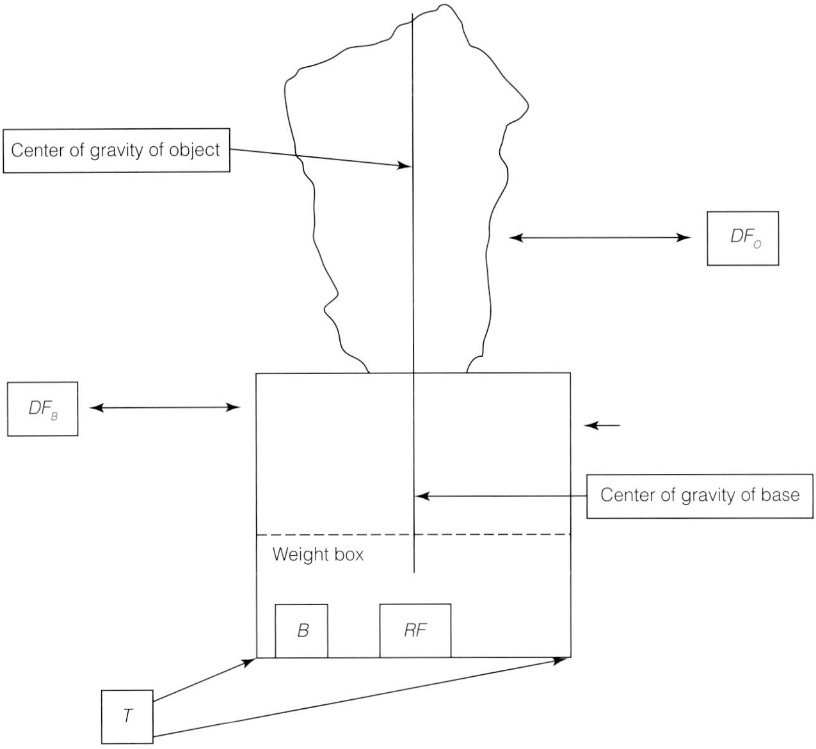

Figure A.1
Diagram of an object assembly (B = minimum amount of ballast needed to prevent toppling, DF_O = driving force to topple an object, DF_B = driving force to topple the base, RF = resisting force to toppling due to weight of base and object, and T = torque points).

The following example is for a monolithic sculpture with dimensions and weight provided. A safety factor expressed as a gravitational risk factor is chosen, as are desired base dimensions and ballast. Then the numbers are plugged into the equation to determine how much ballast, often in the form of lead weights or iron bars, is needed to stabilize the sculpture on the desired base.

Once the base dimensions and ballast needs are determined, the base is constructed with a weight box as shown in figure A.1. On the bottom of the box, between the box and the floor, a sheet of Lennite (ultra high molecular weight polyethylene) is attached.

Let: W_o = weight of object
W_b = weight of base

The minimum coefficient or gravitational risk factor is to be 0.45 g. The preferred value for statuary is 0.6 g. Some critical items may need to go as high as 0.7 g or higher (some Northridge Earthquake readings were higher).

By definition:

DF_O = weight of object x gravitational risk factor
$DF_O = W_o(0.45\ g)$
DF_B = weight of base x gravitational risk factor
$DF_B = W_b(0.45\ g)$

RF = weight of object (W_o) + weight of base (W_b)

Example: A monolithic sculpture (the following example is calculated in English measure)
Object dimensions: H: 86 in., W: 25 in., D: 19 in.
Object weight (W_o): 1500 lb.
Gravitational risk factor: 0.45 g
Dimensions of exhibition base: H: 18 in., W: 36 in., D: 36 in.
Weight of base (W_b): 100 lb.

DF_O = 1500 lb.(0.45 g) = 675 lb./g
DF_B = 100 lb.(0.45 g) = 45 lb./g
RF = 1500 + 100 = 1600 lb.
W_o = 1500 lb.
X = 18 in.

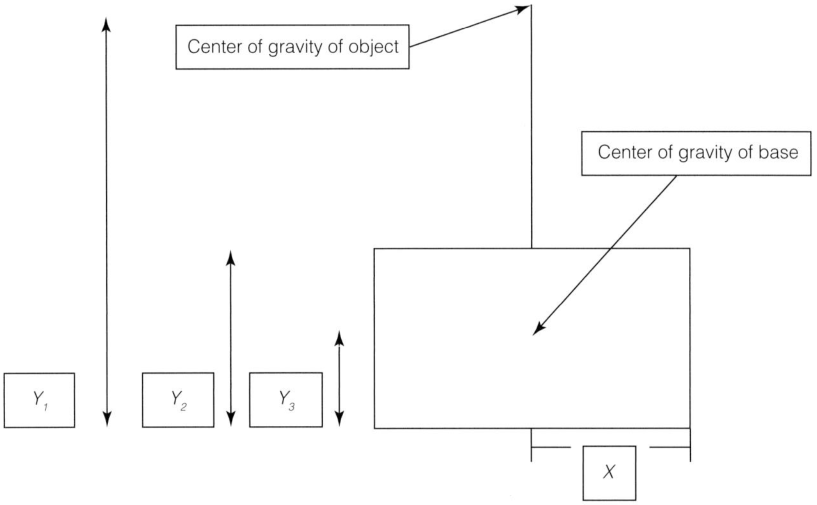

Figure A.2
Simplified diagram of an object assembly (X = one-half of the desired least dimension of the base, Y_1 = height of the center of gravity of the object as installed, Y_2 = the height of the base, and Y_3 = the height of the center of gravity of the base).

Y_1 = 86 in./2 + 18 in. = 43 in. + 18 in. = 61 in.
Y_3 = 18 in./2 = 9 in.

$$0 = \Sigma MT$$

$$DF_O(Y_1) + DF_B(Y_3) = RF(X) + B(X)$$

Or

$$B = \frac{DF_O(Y_1) + DF_B(Y_3) - RF(X)}{X}$$

$$B = \frac{675(61) + 45(9) - 1600(18)}{18} = \frac{41{,}175 + 405 - 28{,}800}{18} = \frac{12{,}780}{18}$$

$$B = 710 \text{ lb.}$$

$$B + RF = 710 \text{ lb.} + 1600 \text{ lb.} = 2310 \text{ lb.} \geq 1500 \text{ lb.}(W_o)$$

Answer: In this example, 710 lb. of ballast would be required to stabilize the object on the base. The ballast weight, added to the combined object and base weights of 1600 lb., results in a total weight of 2310 lb. for the system. The combined weight, which is greater than the object alone, becomes the resisting or restoring force that acts through and lowers the center of mass of the whole system to prevent overturning.

For references to similar physics problems, refer to Weidner and Sells 1965 and Young and Freedman 2004.

Earthquake Protection of Historic House Collections in California State Parks

Nancy Mendez

Karen Hildebrand

Robert Sieger

Introduction

Historic homes are the private residences of individuals and families that have shaped our histories. Their words, actions, relationships, and belongings influence our appreciation of the personal lives of these people and add to our understanding of the broader historic context.

Historic house museums are one of several types of museums identified by such organizations as the American Association of Museums and the International Council of Museums. Of the estimated 17,500 museums in the United States—including art, university, science, and natural history museums—over 2,500 are historic house museums (American Association of Museums; Walker and Graham 1999). Within the 278 properties that constitute the California State Park System, 49 include historic settings/house museums (California State Parks 2000).

The historic house museum is typically filled with objects representing a specific time period and associated with historical people and events. Within the walls of these historic structures is usually an array of objects that ranges from one-of-a-kind works of art to personal items that reflect daily life from a particular period in history. These objects are often kept on permanent display, since they directly reflect the historical period of the structure or site and enhance educational messages and interpretive themes.

Standard museum practices identify several common threats to museum collections. These include pests, light, theft, temperature, humidity, water, fire, and movement. The threat of movement includes both human induced movement, such as the handling of artifacts, and that caused by seismic activity. Seismic risk is clearly as significant a threat as any other—and is, perhaps, even greater, since the event happens suddenly with little or no warning; lasts only seconds or a few minutes; can be completely devastating; and can trigger other threats such as flood, fire, explosion, and mud slides. Because of the magnitude of threat that earthquakes pose, it is paramount that collections consider earthquake damage mitigation as part of their preventive conservation and collections care program.

Many historic house museums are operated by state agencies and local organizations that are typically understaffed and underfunded. When these factors are combined with the high seismic activity in areas such as California, it is imperative that simple and low-cost solutions be developed to improve seismic stability for house museum collections.

This paper discusses earthquake protection strategies that have been used in historic house collections within California State Parks and which have been implemented through a multidisciplinary team approach for developing low-cost and simple solutions. The first part of this paper discusses a seismic survey and pilot mount project at the historic home of Will Rogers. The second part discusses a current museum exhibit project at the historic de la Ossa Adobe, which has been closed since the 1994 Northridge Earthquake. The final section of the paper reviews the need for a multidisciplinary team approach in order to optimize seismic stability solutions.

Case Study: Will Rogers Home and Collections

Will Rogers State Historic Park, the site of the Will Rogers Home and Collections, is located in the Santa Monica Mountains, within a few miles of the Pacific Ocean, in Pacific Palisades, California. The Will Rogers Home began as a simple cabin in 1926 and over the next few years was expanded into the Ranch style house that can be viewed today (fig. 1).

This home of noted humorist, writer, and actor Will Rogers is listed on the National Register of Historic Places. It contains the original furnishings owned by Will Rogers and his family and includes an eclectic collection of gifts from his many admirers, as well as objects Rogers purchased during his travels around the world.

The Rogers home and property have been part of the California State Park System since 1944. Although a few of the objects have received conservation treatment and mounting improvements over the years, the collection had not received an assessment specifically aimed at identifying seismic stability issues.

The Seismic Stability Survey In 1994 staff from the J. Paul Getty Museum's Department of Antiquities Conservation conducted a joint project with curatorial staff at Will Rogers State Historic Park to assess existing mounting problems and recommend simple, inexpensive solutions (Sieger 1994).

In order to assess the mounting problems, a comprehensive survey was conducted for the objects on display in the Rogers Home. A simple form was created to record basic information about an object and its general seismic status. The form included fields for recording object name, location, materials, seismic status, and comments.

Seismic status was generally recorded as either stable or unstable; the "comments" field provided an opportunity to elaborate on specific seismic areas of concern or to recommend specific mounting improvements. The areas of concern fell into four basic categories: (1) objects located too close to the edge of a surface (i.e., table, cabinet, or shelf); (2) objects located too close to other objects (particularly if either or both are fragile pieces); (3) objects located on unstable surfaces; and (4) objects that lacked stability because of high centers of gravity.

From the resulting data, eight objects were selected to receive mounting improvements. The selected objects represented a broad range of materials and mounting problems. The solutions illustrate many different types of

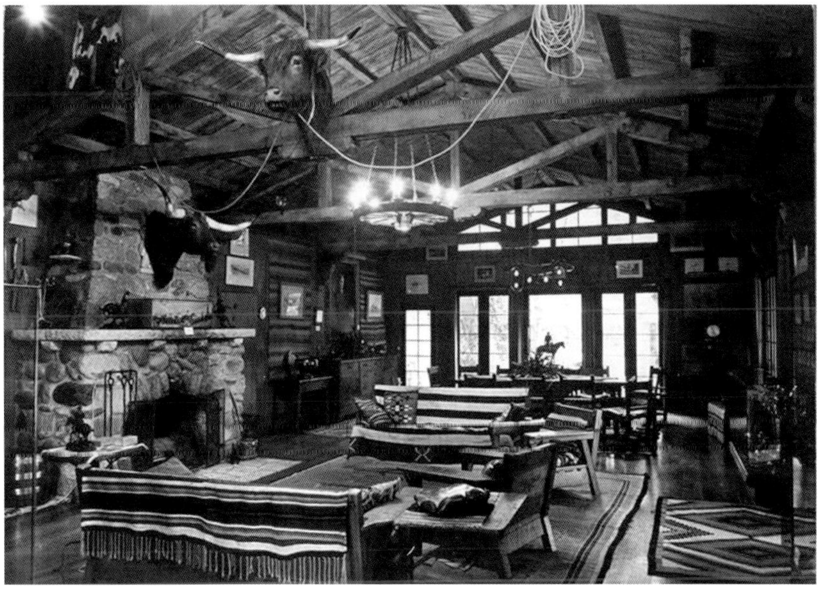

Figure 1
An early photo of the eclectic collection at the Will Rogers Home. Photo: California State Parks, Will Rogers State Historic Park.

mounts, all made with easily available materials and tools. All the mounts were designed to be made by anyone with even a minimal level of technical skills. The section that follows discusses the key stability issues that were identified and reviews the subsequent restraint and mounting improvements that were made.

Low-Cost and Simple Solutions

Repositioning Objects and Minimizing Object Movement: Dining Table Setting The seismic survey indicated that several objects were at risk because of their placement within the historic house museum environment. Typical of house museum vignettes or settings, the Will Rogers Home contains a dining area with dishware and glassware displayed on a dining table. A number of the objects on the table were displayed too close together, posing a threat in the case of even a slight amount of seismic activity. (The force of an impact on an artifact is much greater than the force required to move it, such as an earthquake, because of the concentration of the force on a very small area.) By separating these pieces from one another to the extent possible, the seismic threat was easily reduced.

The dishware and stemmed glassware in this collection have a low height-to-base ratio and did not appear to be in danger of tipping in the event of an earthquake. However, some plates were displayed very close to the table's edge and were more vulnerable to falling with even a small amount of movement in the area. Moving these objects toward the center of the table was another simple way of lessening seismic risk.

Unlike other historic dining table vignettes, the Rogers dining table is not covered with a tablecloth. Previous attempts had been made by park staff to add barriers between the various objects on the table and the table itself, primarily to protect the table's surface. However, during the seismic stability survey, it was determined that acid-free paper and polypropylene barriers actually contributed to increased seismic risk.

In reassessing the need for a barrier to also provide seismic stability for the objects, silicone sheeting was selected to be the best material because of its ability to reduce the amount of sliding of an object placed on top of it. Chemically inert silicone is available in very thin sheets and can be easily cut with a pair of scissors. In this instance, a thickness of 0.8 mm (1⁄32 in.) was used. The silicone sheeting was cut into pads slightly smaller in size than the object bases, so that they were not visible. By carefully repositioning objects (i.e., away from one another and away from the table edge) and by placing silicone pads between objects and the table, seismic risk was reduced while the overall historic character of the dining table display was retained.

Restraining Object Movement: Books Another example of objects that were located too close to the edge of a surface included the books displayed in the Rogers Home. Several hundred books—many personally inscribed to Will Rogers—are displayed on open shelving in the Rogers family library. The bookcases are built into the wall of the historic home and appear structurally stable. However, there are no restraints to keep the books from falling on the floor in the event of seismic activity.

Some key issues—common to most historic homes—needed to be considered while the appropriate restraint method was being developed. These included:

- The restraint method must be visibly unobtrusive to maintain the historic character of the Will Rogers library and home.
- In order to preserve the historic integrity of the Will Rogers Home, which is listed on the National Register of Historic Places, the bookcases and shelves cannot be significantly altered.
- Ongoing preservation needs require access to the books and bookcases for activities such as routine pest inspections and regular cleaning of the books and shelves. A relatively easy way to disassemble and reassemble restraints is needed.

The simple and low-cost solution was to attach small brass screw eyes to each side of the bookshelves, then stretch 17 lb. test monofilament line from eye to eye, tying each end to the screw eyes (fig. 2). Ends were tied with a common knot of two half hitches.

This type of restraint satisfied each of the key concerns. First, the line is barely visible from even a short distance, yet it serves to prevent books from falling off the shelf in the event of an earthquake. Second, each screw eye hole is extremely small in diameter, less than 1.6 mm (1⁄16 in.). And third, during regular preservation maintenance activities, the monofilament line can easily be cut and replaced.

Restraining Object Movement: Musical Instruments on Furnishings A violin in an open case is displayed resting on an overstuffed armchair. The violin was secure in the case, but the case was unrestrained and unstable. Again, 17 lb. test monofilament was used, in this instance to tie the case to the arms of the chair. The line was passed around the hinges of the case and is not visible.

A mandolin is displayed upright on a loose-pillow sofa. It could easily fall to the floor. Again, the monofilament was used to secure the object. The line was passed under the mandolin strings just above the nut and sewn to the back

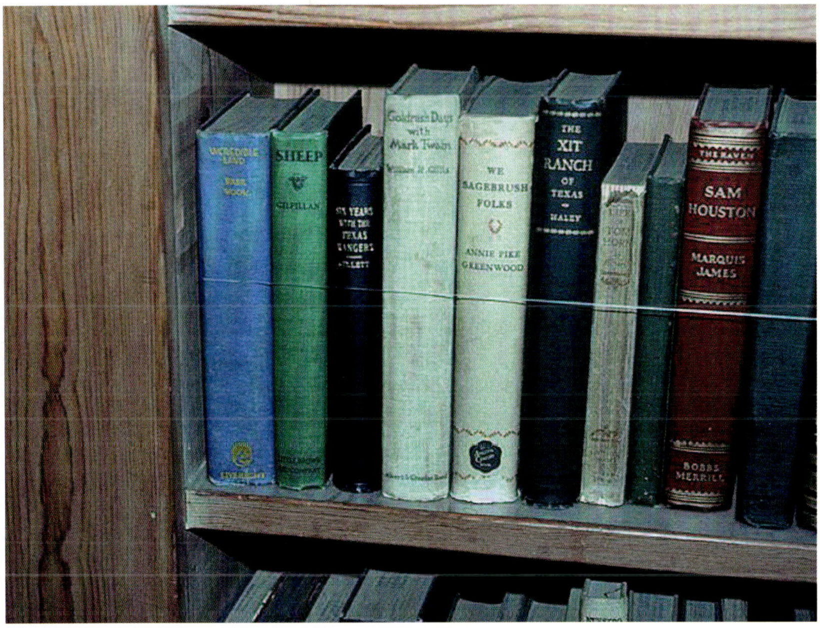

Figure 2
A simple and low-cost solution for securing books: they are restrained with a length of 17 lb. test monofilament line. Photo: Robert Sieger.

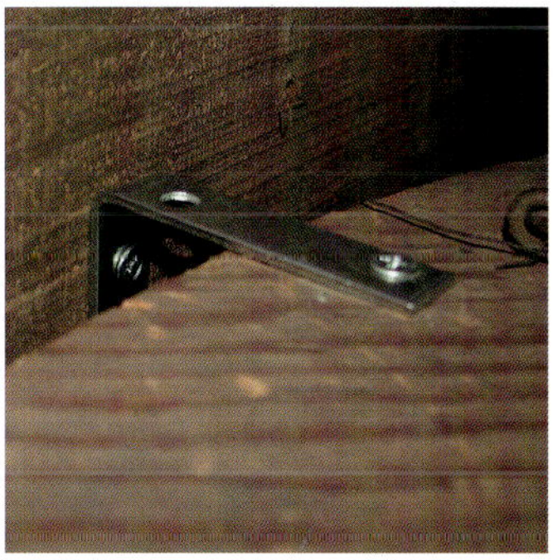

Figure 3
An example of a metal bracket securing a cabinet. California State Parks, Will Rogers State Historic Park. Photo: Robert Sieger.

pillow. A curved needle was used, and care was taken to sew the line between and not through the threads of the pillow fabric. The pillow was then sewn to the upholstered back of the sofa with the same technique and materials.

Restraining Object Movement: Cabinets Among the other objects identified as unstable was a tall oak and glass bow-front cabinet that occupies the southwest corner of the main living room. Past attempts to secure it to the walls are evident. Two common screen door hooks are screwed to the top of the cabinet, and one screw eye remains in the wall. No evident attempt was made to secure the bottom. Since screws had been used on the piece previously, and since the walls in the house have many artifacts nailed, screwed, or stapled to them, permission was given to attach four 2 in. (5 cm) corner brackets with small screws to affix the cabinet to the wall.

First, each bracket was held in place two on the top, two on the bottom—and a light pencil mark was made on the wall for positioning. Next the cabinet was moved out of the corner, and the brackets were screwed into the wall. The cabinet was then moved back into place and screwed tightly to the brackets (fig. 3). The brackets are not visible under normal viewing conditions.

Custom Contour Mounts Located on top of the corner cabinet discussed above is a wooden vase. Turned on a lathe from a solid piece of cypress, the vase is tall, narrow, and unstable. It was determined that this object would be best protected by an acrylic contour-type mount. A ¼ in. (6 mm) thick piece of clear acrylic was cut on a scroll saw to closely match the outline of the vase. The precise negative shape was obtained with a plastic contour gauge. Two small holes were drilled through the contour, one at the top of the mount and the other near its base, through which monofilament was strung to secure the object to the mount. The contoured piece was adhered to an acrylic base, thus forming a kind of exterior spine support for the object (fig. 4). To prevent the

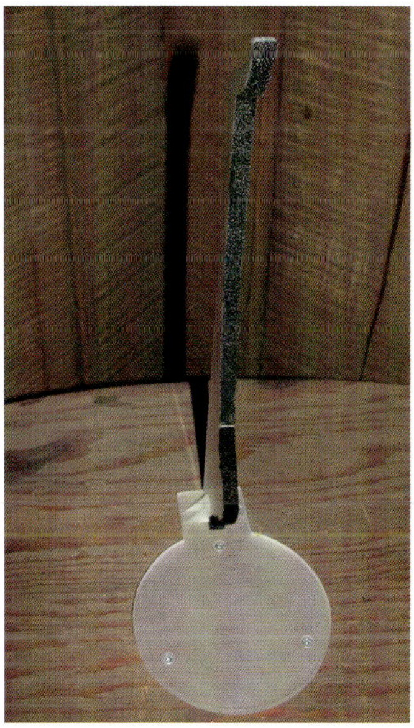

Figure 4
An acrylic contour mount for a vase. California State Parks, Will Rogers State Historic Park. Photo: Robert Sieger.

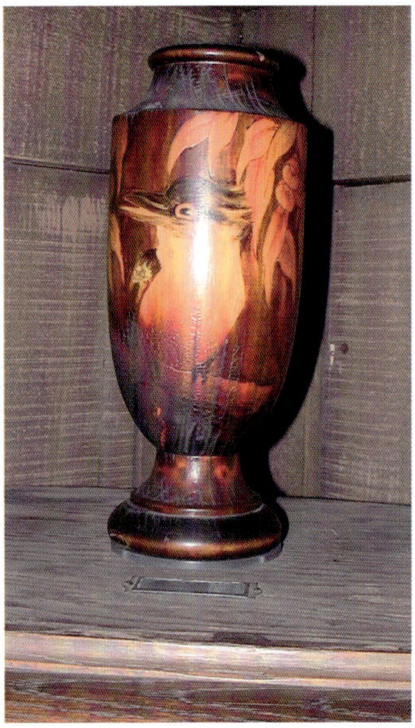

Figure 5
The object attached and installed with a contour mount in place. California State Parks, Will Rogers State Historic Park. Photo: Robert Sieger.

vase from being scratched by the acrylic, a barrier of inert black polyester felt (Deccofelt) was adhered to the contoured edge of the acrylic with Paraloid B-72, an ethyl methacrylate copolymer. The entire mount was then lightly sanded with 320 grit sandpaper to give it an overall matte finish. Next, the mount was screwed to the display deck, in this case the top of the corner cabinet. The object was then placed against the contour spine and tied in two places to firmly hold it against the mount (fig. 5).

This mount was the most complex of the project. To approximate conditions at museums where staff may not be familiar with working with acrylic, the fabrication of this mount was assigned to a student intern who had never made any type of mount before. After receiving basic training on the use of a contour gauge and a scroll saw, the intern was able to construct the mount easily. The finished mount is stable, holds the vase securely, and is barely noticeable.

Simple Mounts: Securing One Historic Object onto Another A common occurrence in historic house museums is the display of one or several historic objects on top of another historic object. During the seismic stability assessment, one example stood out that was in need of attention. A small rectangular wooden table with a single lower shelf is located at the north end of the living room. A small rounded ceramic pot is displayed on this shelf. Although the table is stable, the pot is unstable and could easily roll or slide off the shelf, falling to the wooden floor. A good method of stabilization for an object of this nature is to adhere it directly to the shelf with a synthetic microcrystalline wax. However, to avoid any possibility of staining, some type of barrier must be applied between the wax and the pot as well as between the wax and the shelf.

It was determined that an acrylic sheet would protect the shelf surface and act as a barrier between the shelf and the bottom of the pot. A 1⁄16 in. (1.6 mm) thick clear acrylic sheet was selected and cut to match the size of the shelf. Small holes were drilled in the corners of the acrylic sheet. Next, the sheet was placed on thin, clear protective spacers on the shelf. These spacers were also made of acrylic sheet; these allow air to circulate under the acrylic sheet, reducing the possibility of damage to the shelf by condensation. Monofilament was passed through the holes in the acrylic sheet and tied under the shelf.

A 15% solution in acetone of Paraloid B-72 was applied to three small surface areas on the perimeter of the bottom of the pot and allowed to dry thoroughly. Three small beads (approx. 2–3 mm, or 1⁄8 in., in diameter) of synthetic microcrystalline wax were placed on these prepared areas. The pot was then positioned on the shelf; a slight downward twisting motion was used to flatten the wax beads and adhere the pot to the acrylic sheet covering the shelf.

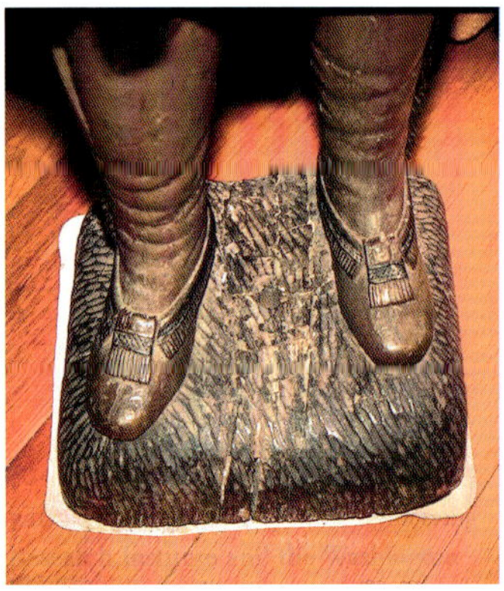

Figure 6
The former mount (a hand towel placed under the object), which was somewhat visually distracting. California State Parks, Will Rogers State Historic Park. Photo: Robert Sieger.

Custom Mounts: Tall Objects with Small Bases A statue of a Native American, carved from a single block of wood, stands next to the dining table. Standing 1.5 m (5 ft.) tall on a narrow base—approximately 35.6 cm (14 in.) square—this statue has a high center of gravity and could easily topple. The old "mount" was simply a terry cloth hand towel placed under the artifact (fig. 6). While this measure may have slightly encouraged the object to slide rather than rock and possibly overturn during an earthquake, it was much too small to be effective.

It was decided to make a new, unobtrusive base large enough to favorably change the aspect ratio of the artifact, thus lowering its center of gravity. This alone would make the artifact much more stable. After measuring the height of the artifact, it was determined that a base width of approximately one-half the height of the sculpture was the largest possible base size aesthetically possible.

A piece of ¾ in. (19 mm) thick exterior plywood was cut to the correct size. Exterior plywood was used because the adhesive used in its manufacture is not formaldehyde based. The thickness of ¾ in. was chosen because it is a standard thickness and does not add to the instability of the object by flexing and acting as a spring. All the edges of the base were rounded over, using a router, to soften the outline. Pads of Teflon sheet, chemically etched on one side to allow adhesion to the plywood, were adhered to the underside of the new base. These would further ensure that the artifact would tend to slide rather than rock in an earthquake. The new base was then painted flat black with acrylic spray paint, which was slightly darker than the object, to make it less obtrusive.

The base of the statue is roughly 15 cm (6 in.) thick and has an existing through hole measuring slightly more than ¾ in. (19 mm) in diameter. This hole was used to attach the new base to the artifact in a completely noninvasive manner. The position of the hole relative to the edges of the artifact was measured, and this position was marked on the new plywood base. A ¾ in. diameter wooden dowel was cut to 12.7 cm (5 in.) in length—a size that would allow for a safe connection from the new base into the preexisting statue hole without being visible (fig. 7).

The top of one end of the dowel was drilled and countersunk to accept a number 8 wood screw. Next, two cuts about 6 cm (1½ in.) deep were made at right angles to each other across the top of the dowel, dividing the upper section of the dowel into four wedge-shaped sections. The other end of the

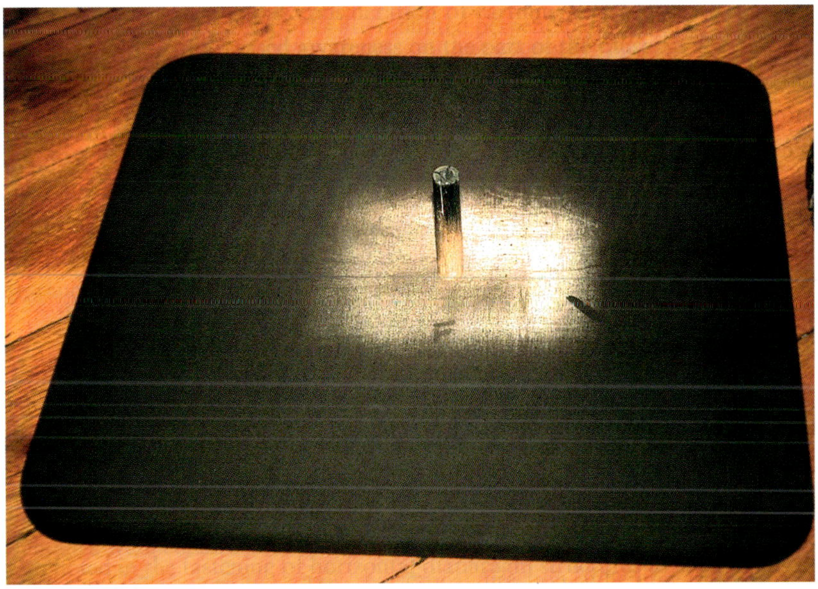

Figure 7
A new mount with its fixing pin in place. California State Parks, Will Rogers State Historic Park. Photo: Robert Sieger.

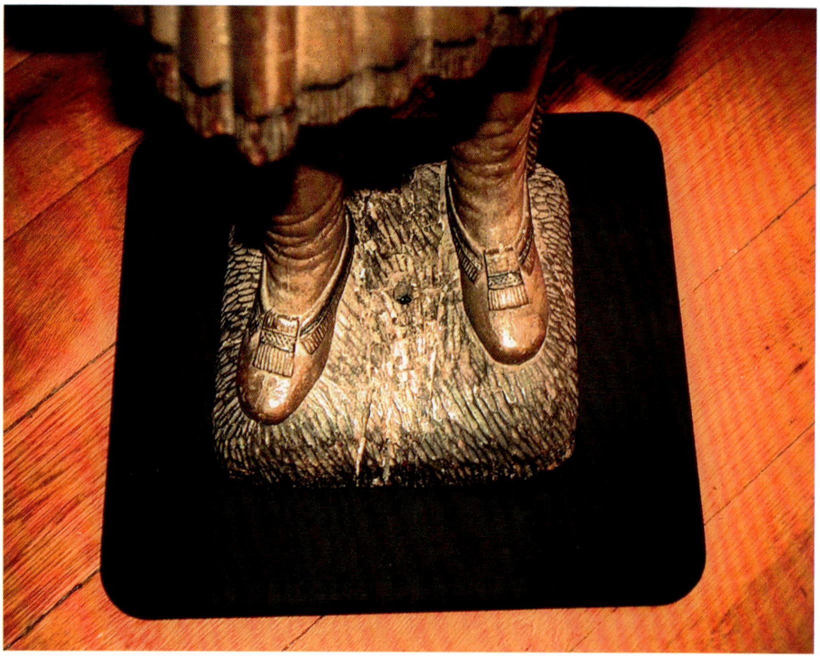

Figure 8
The object installed on the new mount. California State Parks, Will Rogers State Historic Park. Photo: Robert Sieger.

wooden dowel was then glued and screwed to the top of the new base from the bottom of the base.

After the glue had dried, the statue was lowered onto the new base. A number 8 wood screw was then threaded into the top of the dowel. As the screw was tightened, the cut sections of the dowel expanded—firmly gripping the interior of the existing hole—until the new base was securely attached to the statue (fig. 8). Although the top of the dowel was not visible from even a short distance away, a small disk of inert black felt (Deccofelt) was cut and pushed into the hole to completely mask it.

This mount was inexpensive to construct and took only a short time to fabricate. By adding this larger base and carefully securing it to the statue, the center of gravity of the object was significantly lowered, and it is much less likely to fall during seismic activity.

Seismic Threats to Adobe Structures

While earthquakes in California have damaged wooden historic buildings, the impacts have been much more harmful to those built of adobe. Some of our most significant historic buildings are adobe because this traditional construction approach was so predominant in the California missions as well as early secular ranchos and pueblos, particularly in Southern California. Two examples, in addition to the case studies for this paper, may help illuminate the extent of the conservation challenges in the California State Park System.

At Santa Cruz Mission State Historic Park, the state owns the only surviving housing built for neophytes (converted California Indians) in the entire chain of missions in Alta California. It suffered severe damage in the Loma Prieta Earthquake of 1989, which measured 7.1 on the Richter scale. The entire building would almost certainly have collapsed if seismic stabilization had not been in place, although the restoration was not complete at the time.

Pío Pico, the last governor of Mexican California, owned a grand rancho east of Los Angeles. State Parks had opened his adobe home to the public decades before the Whittier Narrows Earthquake of 1987, when it was heavily damaged. Following the earthquake, it took State Parks several years to complete recording, budgeting, and implementing stabilization and rehabilitation, as well as installing completely new interpretive exhibits. During this time, the importance of the surrounding landscape was more fully recognized, and studies have led to the installation of rancho-era orchards, vineyards, and crops to provide a microcosm of the original setting.

Case Study: de la Ossa Adobe and Collections

The second case study is an earthen building: the de la Ossa Adobe is located in Los Encinos State Historic Park, 29 km (18 mi.) northwest of the original Pueblo de Los Angeles. The eight-room building was constructed in 1849 to house the owner's family at the 4,460 acre Rancho El Encino. The adobe evolved to reflect both the changing times and the changing cultural affiliations of its owners. The adobe homes of Hispanic Californios were simple: the de la Ossa Adobe had mud plaster walls finished with whitewash, and *manta* (cotton cloth) suspended above, in lieu of ceilings. French owners in the 1870s repaired earthquake damage and added trims, moldings, and elaborate painted decorations, including decorative finishes that are now being treated by a conservator. Basque owners added large fireplaces in 1909–10 but painted over decorative work (Hildebrand 2006).

The de la Ossa Adobe had been fully furnished and open to the public before the Northridge Earthquake of 1994, which registered 6.7 on the Richter scale. The damage included collapse of an entire gable end wall and two chimneys, as well as cracking of all exterior and interior walls (fig. 9). The rehabilitation is now complete, including stabilization consisting of a continuous bond beam at the top plate level. Presently, the authors of this paper are part of the team working on the interpretive exhibits. Team members include a curator, historians, a historical archaeologist, interpreters, and exhibit consultants. These team members have been involved throughout the steps of the exhibit process—including review of text, graphics, objects, and furniture—to assure that historical accuracy and appropriate object treatments are being achieved.

For example, installations for the new exhibits must take into account the information

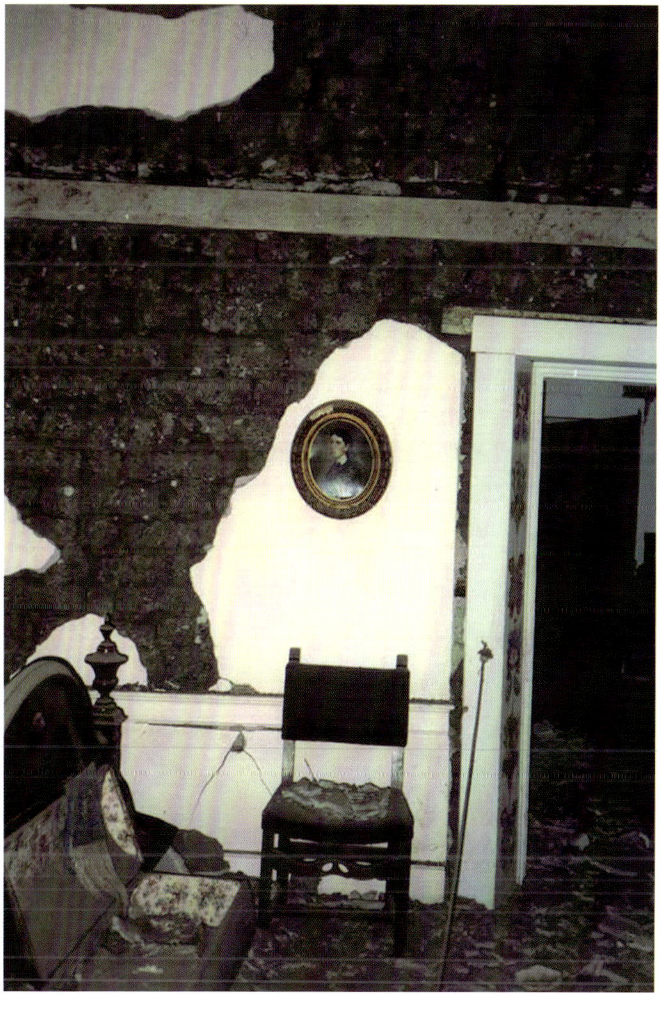

Figure 9
Damage to the interior walls of the de la Ossa Adobe. California State Parks, Los Encinos State Historic Park. Photo: Robert Sieger.

recorded in the fieldwork for the historic structure report. As the consulting exhibit builders presented preliminary plans for the large exhibit panels, we realized that they planned to anchor these from the ceiling. They had gotten the message that the panels could not be anchored to the walls, but they also needed to be given the information that the 1870s ceilings are not sturdy—plus, the historic fabric could not be damaged. The floors, however, are all strong modern replacements. The panels can be attached to the floors to gain needed stability, against both possible visitor impact and future seismic movement.

Low-Cost and Simple Solutions

Mounting Techniques for Framed Works Plans for one of the rooms called for large framed pictures to be hung on the walls. The unreinforced adobe brick walls would not support conventional hangers, but more importantly, invasive procedures should not be allowed. Working with this restriction, the actual details of the building must be discussed and historic precedence considered. The de la Ossa Adobe had crown molding below the original cloth ceilings, and when wooden ceilings were added, crown molding again detailed the top of all walls. Nineteenth-century adobes did not have picture molding—a molding added to walls at about 30–40 cm (12–16 in.) below ceiling level—from which to hang pictures.

Our solution was to use the crown molding to suspend framed works. Because the reinforced bond beam is behind this molding, a secure anchor that can support a heavy picture can be made by drilling into the bond beam just below the ceiling and setting a screw eye from which to hang the hardware. The framed picture can then hang from braided wire that is in full view, in the historical manner.

Anchoring Methods for Lighting Fixtures Suspended from the Historic Ceiling In order to create the Basque owner's 1910 parlor, we needed to suspend a light fixture from the wooden ceiling. The thin beaded board ceiling is nailed to widely spaced joists. To make a stronger anchor for a light fixture, we planned an added 2 × 4 in. (5 × 10 cm) board to run perpendicular to the joists, attached with joist hangers. Its location is to be directly above the location of the lamp so that it can carry the weight. In an earthquake, this could prevent not only the possibility that the lamp might fall but also that the electrical wires might be pulled apart.

Earthquake Protection Training

While this paper offers low-cost and simple solutions to address earthquake protection of historic house museum collections, it must be noted that both the assessment of a collection's seismic stability and the development of appropriate solutions must be undertaken by someone with conservation mount making experience or by someone who has received at least a basic level of training. At minimum, training should include:

1. Methods of Assessing Seismic Stability

- Determining if an object is stable, unstable, or has the potential to become unstable due to factors such as its proximity to edges (i.e., tables, shelves, or cabinets) or proximity to other objects.

- Estimating an object's center of gravity.
- Anticipating object movement.

2. Seismic Stability Solutions

- Determining if the solution will be to allow the object to slide/ move or to be restrained.
- Thorough knowledge of safe mounting practices. One excellent source for this type of information is the Exhibit Conservation Guidelines developed by the National Park Service (Raphael 1994).
- Familiarity with appropriate materials and proper designs that can safely be used near, or in contact with, museum collections. Chemically inert materials must be used that will not stain, rust, discolor, corrode, scratch, bend, cut, or otherwise damage an object.
- An ability to work with other disciplines and specialists to arrive at the appropriate solution.
- A general understanding of the building characteristics of historic homes and how they may affect earthquake protection strategies. Walls, ceilings, and floors must be sturdy enough to support the weight of objects that will potentially be anchored to them.
- Experience in appropriate museum object handling techniques.
- Basic carpentry skills.

The Multidisciplinary Team Approach

Like most other aspects of museum work, the protection of museum collections requires a team approach. As demonstrated in the previous case studies, team members representing multidisciplinary fields are essential in developing and implementing earthquake protection strategies for historic house collections. Depending on the extent of the earthquake protection needs, on the condition of the collections, and on any challenges presented by the historic home, team members may need to include:

- a curator or someone familiar with the museum objects;
- a cultural resource specialist or someone with an understanding of the historic home;
- a historian or someone who knows the historical context of the objects and their relationship to the people, events, and activities associated with the historic property;
- a mount maker or someone with the technical skills to design and fabricate appropriate mounts;
- a museum custodian or technician or someone responsible for the regular care and upkeep of the museum objects and the historic house environment;
- a conservator, particularly if vulnerable or problematic objects are involved.

By utilizing a multidisciplinary team approach to address earthquake protection issues, simple and low-cost solutions can be achieved that optimize both the seismic stability of museum collections and the integrity of the historic house museum.

References

American Association of Museums
http://www.aam-us.org/aboutmuseums/abc.cfm, "Museums FAQ: 1. How many museums are there in the United States?"

California State Parks
2000 *Park-to-Park Index* (Sacramento).

Hildebrand, Karen
2006 *Historic Structure Report: The de la Ossa Adobe* (San Diego).

Raphael, Toby
1994 "Design and Fabrication of Conservation Mounts," *Exhibit Conservation Guidelines* (Washington, D.C.).

Sieger, Robert
1994 "Will Rogers State Historical Park Pilot Mount Project," Park History Files, California State Parks, Southern Service Center, San Diego.

Walker, Patricia Chambers, and Thomas Graham
1999 *Directory of Historic House Museums in the United States* (Lanham, Md.).

Protecting Temporary Exhibitions from Earthquakes: *Jean-Antoine Houdon: Sculptor of the Enlightenment*

Brian Considine

Ziyad Durón

Introduction

The J. Paul Getty Museum has had a program for the seismic protection of its collections for more than twenty years. During that time, the Museum has undertaken several collaborative projects with the engineering departments at several local universities, as well as extensive work with consulting engineers.[1] Additionally, the two locations and buildings housing the Getty Museum (the Getty Center and the Getty Villa) have completed response spectra studies for each site. These studies have informed years of work by the Department of Preparations, the Department of Antiquities Conservation, and the Department of Decorative Arts and Sculpture Conservation to design and fabricate appropriate showcases, pedestals, and specific seismic hazard mitigation mounts for the objects in the Museum's collection.

The opening of the Getty Center in 1997 saw the beginning of an ambitious program of temporary exhibitions featuring works of art borrowed from other institutions. Prior to the renovation of the original Getty Villa in Malibu, such loan exhibitions were not possible due to spatial constraints. This paper will discuss the challenges of protecting works of art loaned to the Getty by other museums for temporary exhibitions. A specific case study of the mount making efforts for one such exhibition, *Jean-Antoine Houdon: Sculptor of the Enlightenment*, will serve to illustrate the Museum's efforts in seismic damage mitigation.

Jean-Antoine Houdon: Sculptor of the Enlightenment

The Houdon exhibition was on view at the Getty Center from November 4, 2003, to January 25, 2004, and included sixty-nine pieces of sculpture, mostly busts in bronze, marble, terracotta, and plaster. Objects ranged from small busts to monumental sculpture and weighed from 5 to 1650 lb. (2.3–748 kg). As is often the case, the Getty had coorganized the exhibition with other institutions; in this instance the Museum's partners were the National Gallery of Art in Washington DC and the Musée National des Châteaux de Versailles et de Trianon in France. Traveling exhibitions such as the one focused on the works of Jean-Antoine Houdon follow a schedule of approximately three months of display, with two months between the closing at one museum and the opening at the next venue. The two months of down

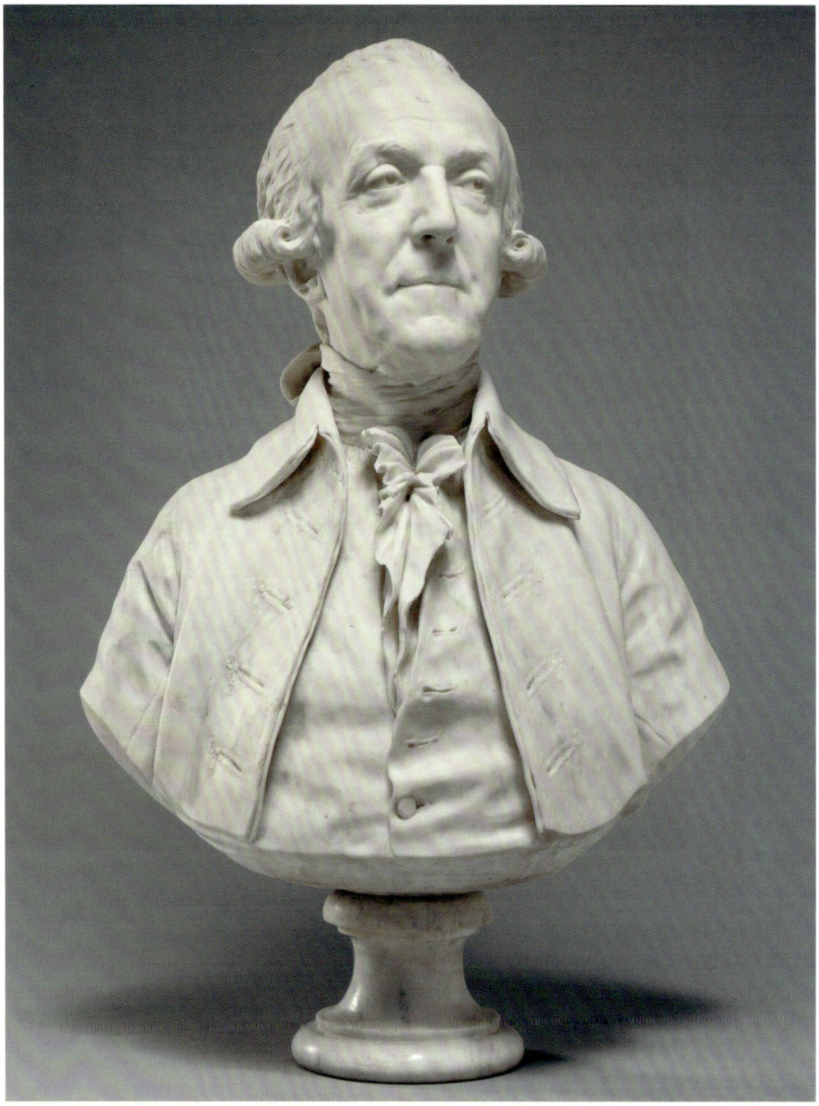

Figure 1
Jean-Antoine Houdon
French, 1741–1828
Bust of Marie-Sébastien-Charles-François Fontaine de Biré, 1785
Marble, overall height: 82.3 cm (32⅜ in.); height of bust without socle: 67.7 cm (26⅝ in.)
Los Angeles, J. Paul Getty Museum, 2003.102.

time in a museum's temporary exhibition space are spent deinstalling the closing exhibition; packing and shipping those works of art; doing the construction, painting, and lighting to ready the space; and receiving and installing the objects for the next exhibition. For museums with seismic risks, however, the challenge is to design, fabricate, and install mounts resistant to earthquake forces for objects that are not immediately accessible, since they may have not yet arrived on loan. The fabrication of special mounts must be completed within a narrow window of opportunity, and extra time is rarely allotted for this important work.

Planning the mounts for the works by Houdon was particularly challenging since there were a number of factors that made the sculptures particularly vulnerable to seismic activity: they were mostly busts on relatively small socles, or sculpted bases, displayed above eye level, and they had very high centers of gravity with unfavorable aspect ratios (the aspect ratio is the ratio of the height of the assembly to the width of its base; see fig. 1). Compounding that

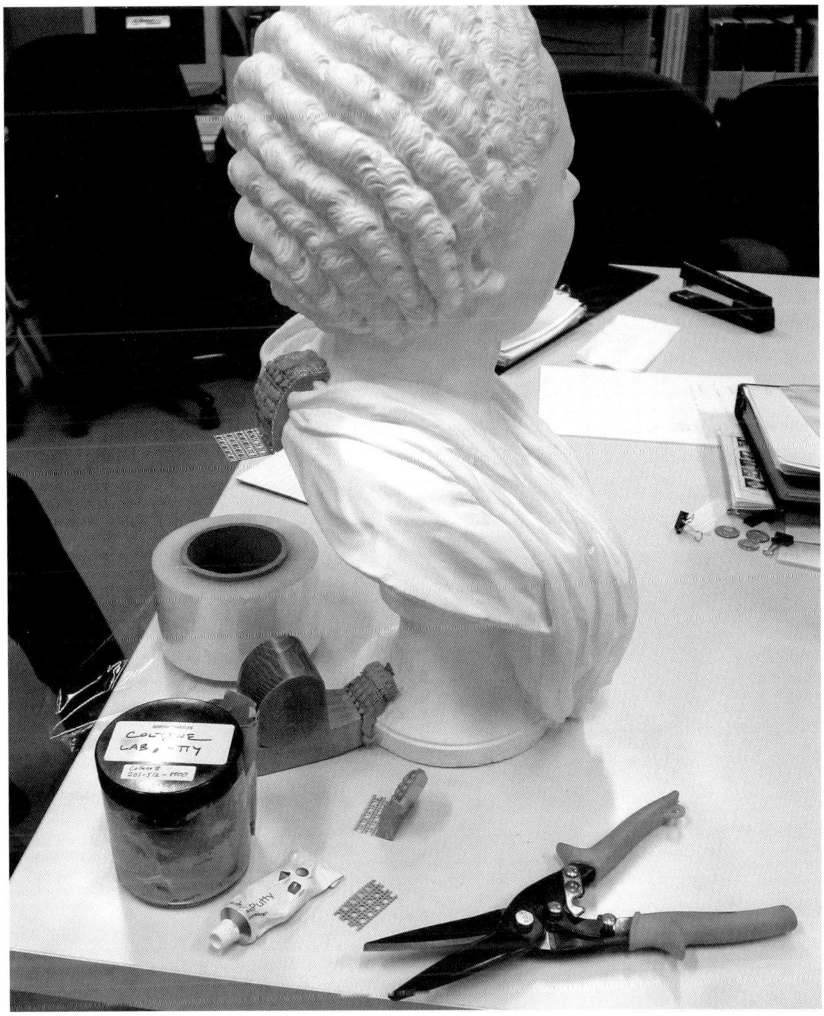

Figure 2
Coltène epoxy putty backed with aluminum mesh, molded to the socle at the base of the bust and to the upper back segment of the drapery. The marble is protected from contact with the putty by a layer of plastic sheeting.

risk was the fact that the busts and the socles were almost always separate pieces, making a solid connection between them of critical importance. Last, there was a variety of materials making up the objects, ranging from bronze and marble to plaster and terracotta, all having different material strength characteristics.

Meeting the exhibition deadline required that the mounts be made before the arrival of the artworks at the museum. To that end, Museum staff visited each of the lending institutions to gather the necessary information and measurements. Using a portable metal detector similar to those used by airport security agents, the conservation team was able to confirm one of the most critical pieces of information: in every instance the busts were firmly joined to their socles with a metal pin. Extensive photographs were also taken, dimensions were recorded, and impressions were taken of the socle profiles with Coltène, a two-part silicone epoxy, backed in these instances with a piece of aluminum mesh (fig. 2).[2] These impressions would be used to manufacture bronze clips to secure the socles to the tops of the pedestals.

The consent to this study by the lending curators was vital, and their understanding of necessary precautions important. Getty Museum staff were careful to explain their plans and needs in writing ahead of the visits and to repeat those explanations in person while working at the various lending institutions. There were, however, a great many additional details still to be worked out. With a spreadsheet of all the data and notebooks of photographs, the Getty Museum turned to Ziyad Durón, professor of engineering at Harvey Mudd College in Claremont, California, who joined the project team.

The Engineering Studies

The design of effective earthquake-resistant structures is a well-known effort within the engineering profession; however, the practical design of earthquake-resistant hardware aimed at protecting priceless artifacts is less well understood. The problem stems from the unknown risks associated with the expected seismic event and with the often fragile condition of the artifacts. Working with Prof. Durón, the Getty Museum sought to better understand the nature of the seismic risk from a structural standpoint before embarking on the design of hardware (mounts) for each artifact.

Environmental Study

A study of the Special Exhibitions Pavilion at the Getty Center was undertaken to gain a better understanding of the response behavior of this structure during a seismic event. Results from the survey were used to determine the relationship between the response of the building to potential excitation frequencies caused by an earthquake and the response of the gallery floor, within the Pavilion, to those forces. The Special Exhibitions Pavilion floor response spectra were originally developed by Harza Consulting Engineers in 1997. That study provided estimates of floor accelerations that might result from a seismic event at the site of the Getty Center.

Data reflecting the expected floor response levels is used to guide the appropriate design of pedestals and display configurations. These designs are meant to assure that the layout and exhibition furniture did not adversely respond to the peak accelerations of the building and the floors. Without this information, damage to the works of art could result, since the motion at the base of the pedestal (at the floor level) is easily amplified at the top of the pedestal and may cause the artifact-pedestal assembly to overturn.

To assess the amount of amplification that could be present, a response survey of the Special Exhibitions Pavilion was conducted in which actual floor responses were acquired for the lower and upper floors. Using these measurements, an assessment was made of the potential overlap of critical frequencies: those that represented the natural period of the building as well as of the floors in the building, and those frequencies that are likely to be contained in an expected earthquake.

An ambient testing technique was used to determine the critical frequencies of the Pavilion. The advantage of this approach is that the engineer does not have to induce an excitation of the structure in order to measure its natural frequency. Instead, the excitation is provided by ambient, environmentally induced excitation, such as wind load.[3]

A sample acceleration measurement acquired during the ambient survey of the Special Exhibitions Pavilion on June 2, 2003, is shown in figure 3. The response was acquired at an area west of the Pavilion entry doors. The data

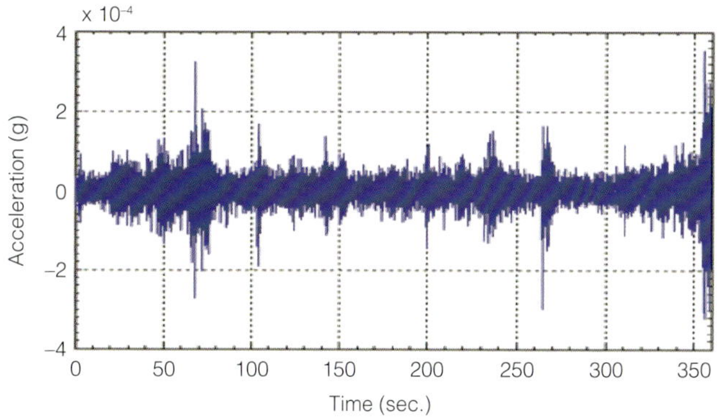

Figure 3
Sample floor response acquired at the Special Exhibitions Pavilion.

are typical of the quality and type of information acquired during the survey at all locations monitored.

The measured response was then analyzed to determine frequency content. Examples of high energy levels are indicated by the large peaks shown in figure 4 in the vicinity of 1, 5, and 17 Hz.

The intent of the survey was to determine whether any measurable responses of the building would contain frequencies that overlapped with the predicted floor response spectra. That is to say, to determine if the response of the building might match the natural period of the floors. Such a situation. if not planned for and mitigated, would cause an amplification of the floor response and endanger the collections. After a review of all the measured responses acquired during the survey, the fundamental frequencies (natural periods) for the galleries were determined as follows:

First resonant frequency	1.22 Hz
Second resonant frequency	1.79 Hz

Figure 5 shows the floor spectra for the Special Exhibitions Pavilion gallery. The red lines indicate the potential overlap of the frequencies determined from the ambient survey for the Pavilion. As shown, the condition does

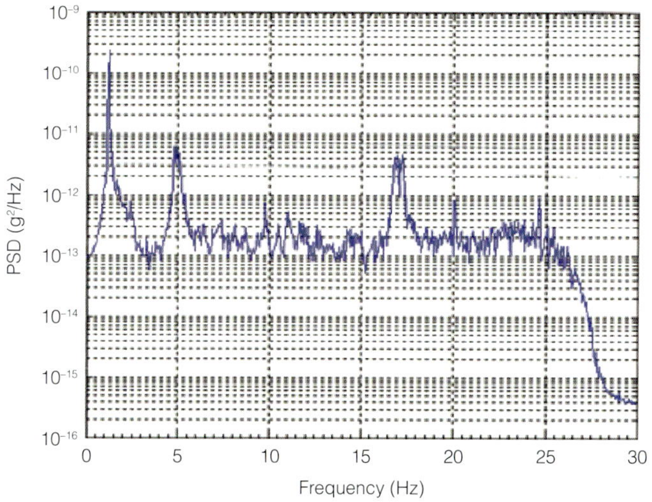

Figure 4
Power spectral density (PSD) at the north gallery location on the second floor of the Special Exhibitions Pavilion. The large peak on the left of the figure corresponds to the fundamental resonance of the north gallery.

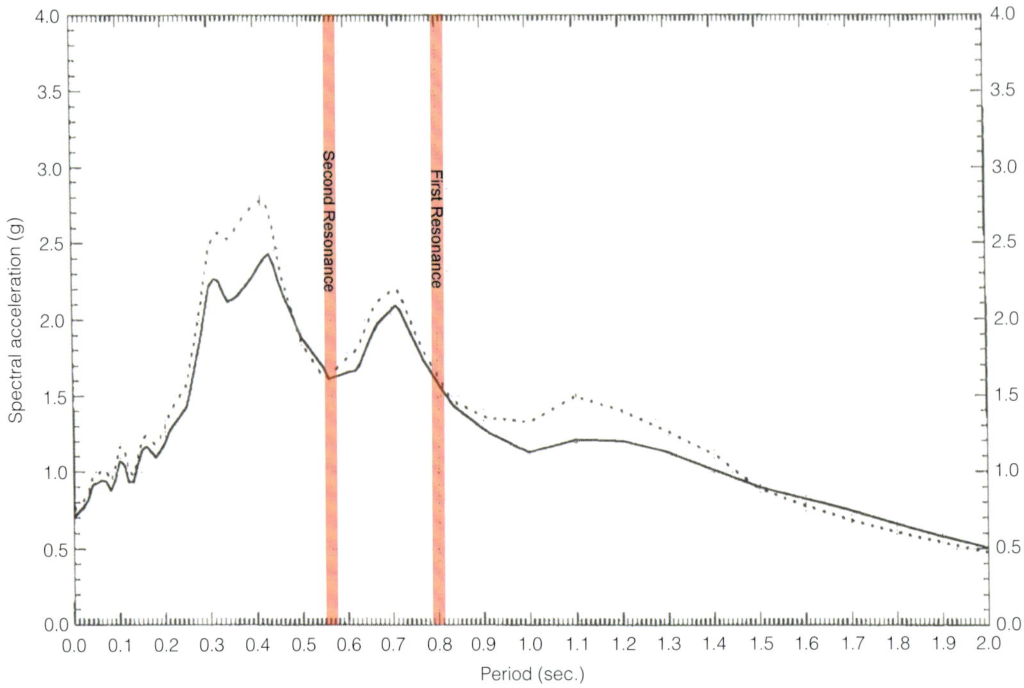

Figure 5
Predicted floor response spectra at an area west of the entry doors of the Special Exhibitions Pavilion. The vertical red lines indicate the periods associated with the first and second resonant frequencies identified from the ambient survey (see fig. 3). A 5% damping is included.

not present a threat, since the resonant frequency for the floor does not overlap with the actual measured behavior of the Pavilion structure. This finding aided the subsequent design efforts, which included the development of criteria to identify artifacts in need of base isolation, specifications of the isolator design, and the design of mounts for securing objects to pedestals.

The Pedestals

The team eventually designed three different types of pedestals for the exhibition: those to be bolted to the structural slab beneath the floor, those that were not attached but were designed to slide during a seismic event, and those with an internal base isolator.

The base isolated pedestals (fig. 6) had been designed by Bruce Metro, the head of the Museum's Preparations Department, and tested at Anco Engineering in 1991 (fig. 7).[4] These tests were conducted under the direction of Paul Burridge, a consulting engineer from Lindvall Richter Benuska Associates in Los Angeles. The isolated pedestals for the Houdon exhibition (fig. 8) stood on ball bearings that allowed for multidirectional lateral movement. A cross brace was internally mounted at the midpoint of the pedestal's height (fig. 9). The cross brace was then connected to the floor with high-strength cable linked to a spring. As the pedestal displaced in the lateral direction, the spring and cable attached to the floor and to the midpoint of the pedestal kept the pedestal from uplifting and overturning. The spring offered resistance to lateral displacement and a returning force to the pedestal. Prof. Durón specified the types of springs used based on the dimensions of the pedestals, the centers of gravity of the objects, and the objects' various weights (fig. 10). The connecting cable's excursion from vertical during the pedestal's displacement was accommodated by the circular opening cut into the bottom of the pedestal, the edge of which

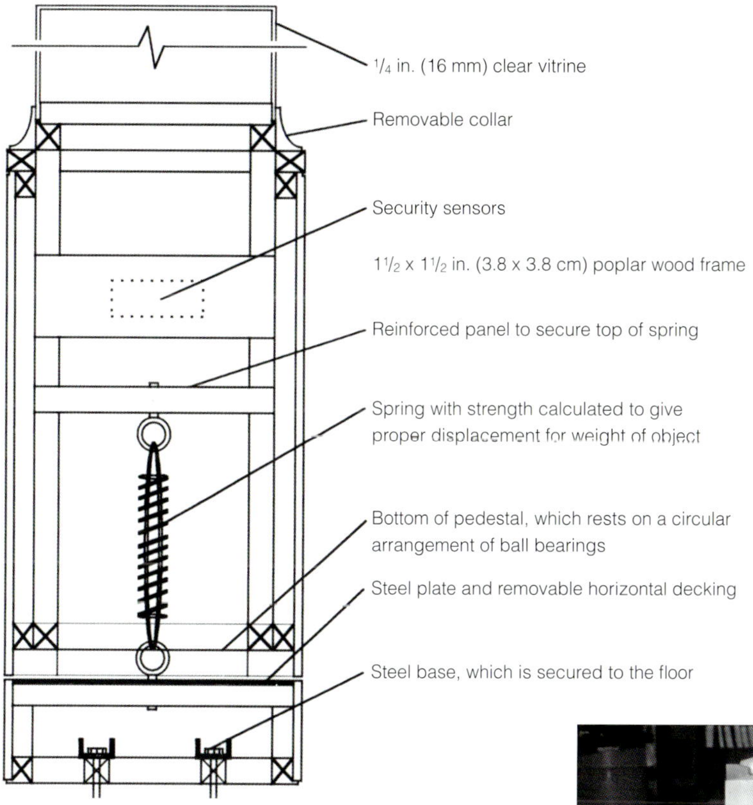

Figure 6
Plan for the construction of the pedestal with isolation mechanism.

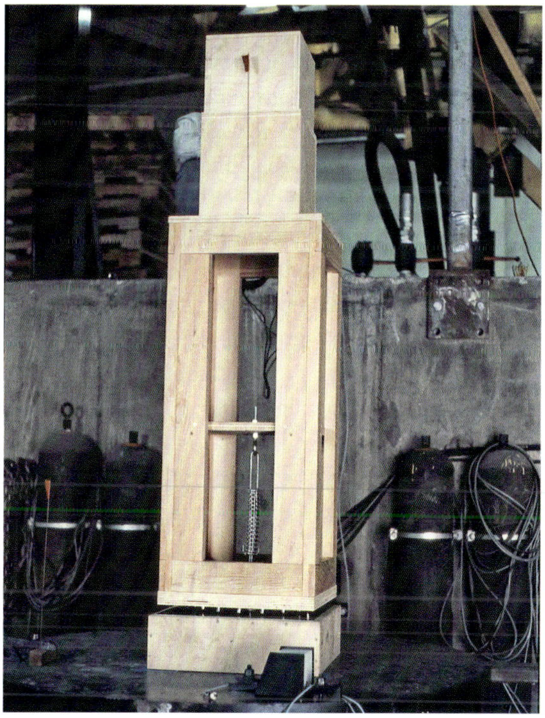

Figure 7
Shake table testing of the pedestal with isolation mechanism at Anco Engineering in 1991.

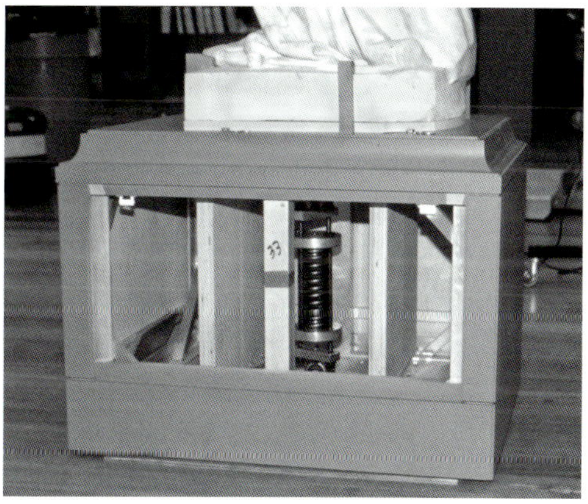

Figure 8
Pedestal with isolator mechanism installed in the Houdon exhibition.

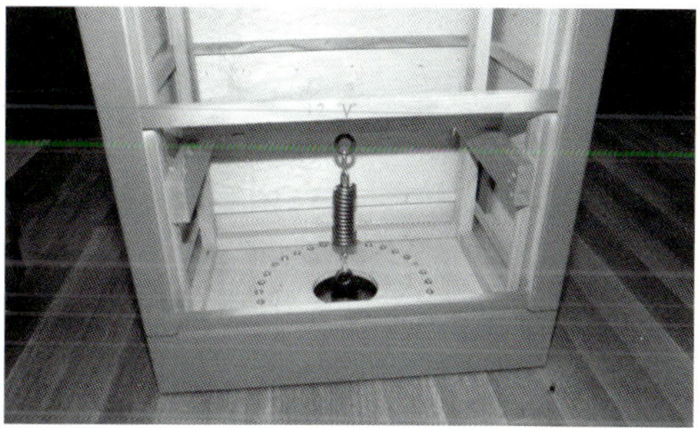

Figure 9
Detail of the isolation mechanism inside the pedestal, showing the resistance spring attached with eyebolts between the floor and the pedestal cross brace.

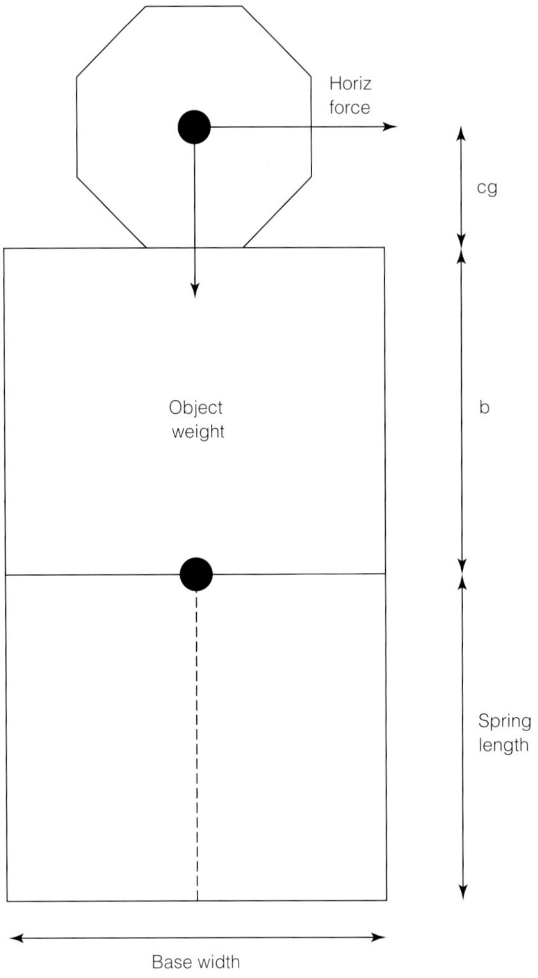

Figure 10
Schematic diagram of the object and pedestal assembly (cg = center of gravity; b = base or pedestal height).

was fitted with a Sorbothane donut-shaped pad that allows for some degree of damping in case the cable comes into contact with the edge of the hole if the isolator reaches its displacement maximum.[5]

Those pedestals that were not isolated but were designed to slide without limiting mechanisms were reserved for situations where the object and display configuration provided a sufficiently low center of gravity and a favorable base-to-height aspect ratio (fig. 11). Fortunately, the exhibition's designer had specified a low height for the large pedestals, and thus they had low centers of gravity and favorable aspect ratios. The undersides of these pedestals were fitted with Teflon skids to allow for sliding. Lead bricks were placed inside the pedestal's lower structure for ballast, to further lower the center of gravity.

The Clips

Regardless of whether the pedestals were anchored firmly to the gallery floor, were allowed to slide, or were isolated, it was paramount that the objects themselves be firmly anchored. Anchoring was achieved through the production and strategic installation of custom-fit clips that mechanically clamped the objects to the pedestals.

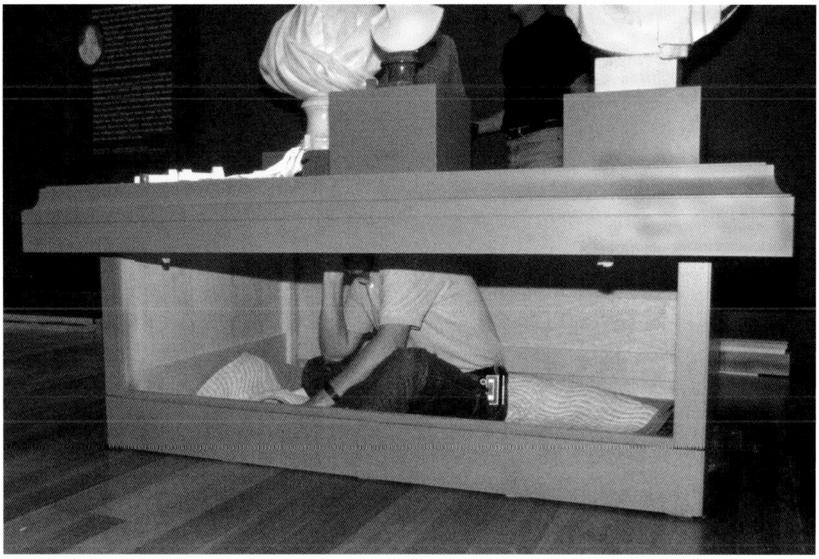

Figure 11
Sliding pedestal design.

The conservation team worked with Prof. Durón to determine both the size and the number of clips per object. The production of the clips was based on impressions taken from the outer profile of the socles. These impressions were achieved by molding Coltène epoxy putty to the appropriate outer sections of the socles (whose surfaces had been protected with plastic sheeting). The hardened Coltène epoxy putty mold was then used to produce a plaster positive cast. During the production of a plaster positive from the epoxy mold, a thin sheet of wax was added to the interior of the Coltène mold to compensate for the Deccofelt[6] pad that would be used between each clip and the surface of the socle of the artwork. Several sizes of L-shaped clips, based on the plaster profiles, were then cast in silicone bronze.[7] Phillyseal[8] epoxy putty was used to fill the gap between the right angle of the clip and the profile of the socle (fig. 12). The face of the cured epoxy that was to come into contact with the object was then lined with Deccofelt to avoid any abrasion of the object's surface.

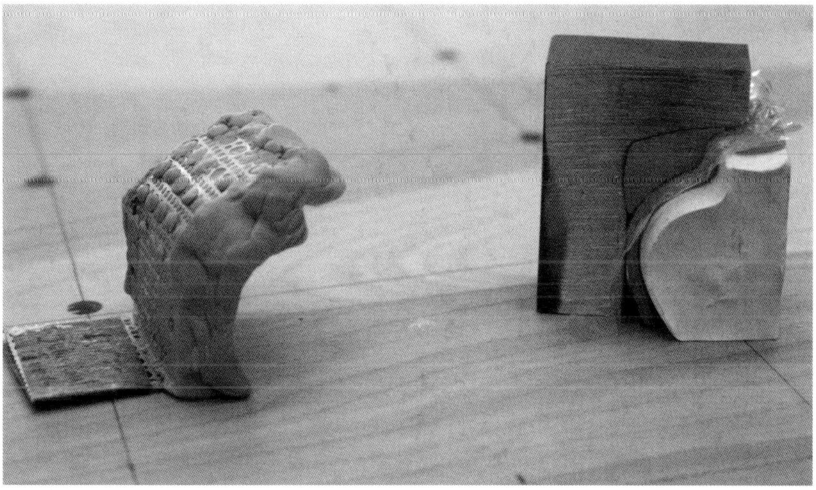

Figure 12
Model of a clip showing a casting of Phillyseal used to fill the gap between the clip and the sculpture.

The bottoms of the clips were tapped to receive bolts protruding through holes in the tops of the pedestals. The pedestal design included a removable panel to allow for access to the pedestal interior and anchors.

The potential failure of the socle at the point of contact with the clips was one of the issues that Prof. Durón studied. Should sufficient force be exerted on the object assembly, fracture of the object would be likely, since the clips themselves are made of a considerably stronger material than the objects. Prof. Durón recommended the number of clips to use for each object. In some instances—based on conservative assumptions pertaining to cracking and actual material strengths—the number of clips suggested by Prof. Durón was quite high. In the end, however, aesthetic considerations led to a compromise, and a maximum of four clips per object was used.

The Installation

Guided by a chart developed by Prof. Durón which listed the objects that required base isolation, as well as by a study of the objects during their installation at the National Gallery (before they came to the Getty Museum), nineteen base isolated pedestals and two sliding group pedestals were planned. When the busts, many of which weighed in excess of 250 lb. (113.4 kg), were attached to the pedestals, the lateral displacement worked well. However, it was found that the overturning, or tipping moment, for some of the assemblies was such that if the pedestals displaced to their limits, the risk of tipping was high. As a result the team decided to bolt seven of the nineteen securely to the floor. The size of the pedestal footprint and the weight of the objects were determining factors in choosing the approach to installation. Those with large footprints and lower weights performed well, while alternative methods were needed for those with small footprints and high centers of gravity.

After the installation of the objects, there was no observed resistance to rotation on the isolated pedestals, so blocks of wax were put under the corners, so that visitors would not see that the pedestals moved.

Conclusion

The *Jean-Antoine Houdon: Sculptor of the Enlightenment* exhibition was both beautiful and popular with the Museum's visitors. There was, however, no seismic event during the run of the exhibition, leaving the project team to make only a highly educated guess about how the installations would have fared had there been an earthquake. The Museum's collaboration with Prof. Durón was very fruitful: the Museum staff involved in the project learned a great deal, and the engineering students at Harvey Mudd College were exposed to problems not usually encountered in their work or studies. The base-isolated pedestals offered considerable protection in the horizontal axis, but at the expense of vertical anchoring. Pedestals with small footprints displaying heavy artifacts presented the greatest challenge. One solution may require the construction of pedestals made of lightweight, high-strength composites that would be firmly anchored and would minimize interactions with the heavy artifacts on display. The Getty Museum will build on this work for an upcoming exhibition of marble sculpture.

Notes

1. The seismic studies that have been performed are: Lindvall, Richter and Associates, "Geologic and Seismic Safety Evaluation of the Main Museum Building," Department of Antiquities Conservation, J. Paul Getty Museum, Malibu, Calif., 1984; M. S. Agbabian, S. F. Masri, and R. L. Nigbor, "Evaluation of Seismic Mitigation Measures for Art Objects," Center for Research in Earthquake and Construction Engineering, Department of Civil Engineering, University of Southern California, Aug. 31, 1989; Woodward and Clyde, "Brentwood Seismic Criteria," Oct. 1989; Robert Englekirk, "Seismic Design Challenges/Brentwood," Robert Englekirk Consulting Structural Engineers, Dec. 1989; M. S. Agbabian, S. F. Masri, and R. L. Nigbor, *Evaluation of Seismic Mitigation Measures for Art Objects*, Center for Research in Earthquake and Construction Engineering, Department of Civil Engineering, University of Southern California, Jan. 1990; C. Eric Lindvall, "Seismic Criteria for Brentwood Museum Contents," Lindvall Richter Benuska Associates, Nov. 15, 1991; Paul B. Burridge, "Horizontal Floor Response Spectra for Special Exhibits," Harza Consulting Engineers, Nov. 25, 1994; John Ash Group, "Decorative Arts Galleries Wall Framing Record Documentation," July 1996; "URS Site Specific Response Spectra and Time Histories for the Getty Center," May 25, 2006.

2. Coltène Lab-Putty, Coltène/Whaledent Inc., 750 Corporate Dr., Mahwah, NJ, 07430.

3. "Ambient Vibration Monitoring at the Getty Museum Special Exhibits Gallery," technical report provided by Z. H. Durón and Daniel Sutoyo, Harvey Mudd College, Claremont, California, June 11, 2003. On file at the Department of Decorative Arts and Sculpture Conservation, J. Paul Getty Museum, Los Angeles.

4. Report on file in the Department of Decorative Arts and Sculpture Conservation: Robert S. Keowne, Project Manager, Anco Engineering Inc., 9937 Jefferson Boulevard, Culver City, CA, 90232.

5. Sorbothane Inc., 2144 State Route 59, Kent, OH, 44240.

6. Deccofelt, 555 South Vermont Ave., Glendora, CA, 91741.

7. Decker Studios, 7400 Ethel Ave., North Hollywood, CA, 91605.

8. Phillyseal R sealing and fairing putty, ITW Philadelphia Resins, 130 Commerce Dr., Montgomeryville, PA, 18936.

Performance Expectations for Microcrystalline Waxes for the Seismic Protection of Art Objects

Anne Crowley

Debra F. Laefer

Background

Use of microcrystalline waxes for the protection of ceramic art objects from seismic events is an inexpensive and relatively popular technique in museum exhibition practice (fig. 1). Unfortunately, because of the high porosity of some ceramics and the fragility of their glazes and paints, the surfaces of many art objects may be vulnerable to damage from the microcrystalline wax application. Thus, a conservative approach is needed—applying only as much as is actually required to resist predicted levels of ground motion and transmitted forces caused by an earthquake's movement. Determining the appropriate and most effective quantity of wax, as well as verifying the best application technique (e.g., hot versus cold), is not clearly established by industrial standards, by museum conservation standards, or by product-oriented guidance. How much wax to use for specific sizes and weights of objects is left to empirical knowledge, judgment, and a good deal of guesswork. While sometimes reliable, these approaches can lead to the application of a greater amount of wax than needed (resulting in unnecessary risk to the object) or too little wax with respect to the object's mass and the anticipated earthquake threat (leading to an increase in the potential for earthquake damage).

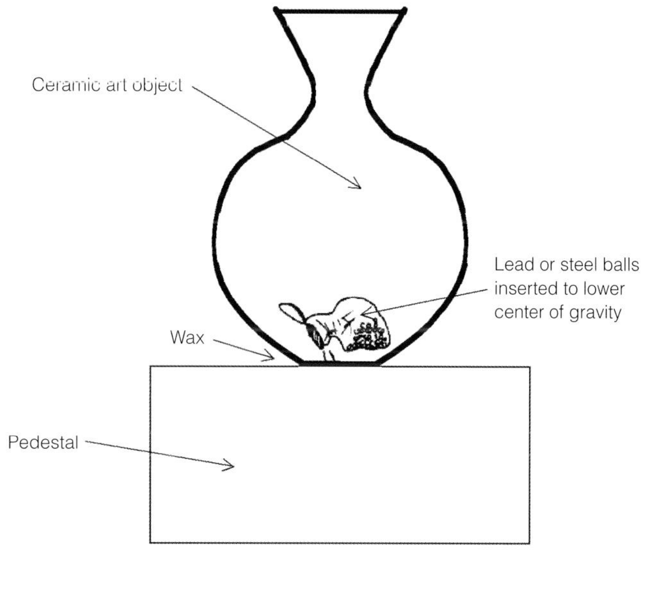

Figure 1
Typical arrangement for the application of microcrystalline wax to an art object. Drawing: Anne Crowley and Debra Laefer.

This ongoing study has begun to develop some performance expectations for microcrystalline waxes and suggests a number of application guidelines, based on chemical microstructure and physical tests.

Testing

The initial testing regimen consisting of 100 tensile and 215 shear tests established the range of physical performance expectations, in terms of minimum and maximum capacities, as well as consistency. Six waxes were tested: paraffin, beeswax, and four microcrystalline waxes (Multiwax, Secure Wax, Quake Hold, and Earthquake).

Conservators usually apply microcrystalline waxes cold, often with a protective coating (an acrylic resin, Paraloid B-72) applied to the specific area of the object's surface, where the wax will be in contact. In the tensile and shear tests, investigations into the impact of application method on capacity were conducted. Samples were tested that had been applied by a hot method (the wax was liquid when applied) and by a cold method, in which the wax was applied at approximately room temperature. The application methods were tested with and without the protective coating applied to the object surface.

Nuclear magnetic resonance (NMR) and scanning electron microscopy (SEM) were conducted on the waxes tested to arrive at a greater knowledge of the relationship between crystallinity of the wax (the extent to which a three-dimensional order exists on the atomic level of dimensions, based on definite and ordered chemical and geometrical structure), branching (a type of structure), and predicted shear and tensile capacities. To better understand the results and to find a simple test that could serve as a reliable predictor of tensile and shear performance, further experiments were conducted, related to relative hardness (through needle penetration tests), contraction, density, softening point, and melting point.

Results

Candle waxes were found to be inappropriate because of their brittle failure mechanisms (figs. 2 and 3a–c). Results of the tensile evaluations indicated that the capacities of the waxes ranged from 129–277 kPa hot applied and 107–208 kPa cold applied. The protective coating increased the capacity of the hot-applied samples (by 51%); however, its impact on compacted samples was inconsistent (table 1).

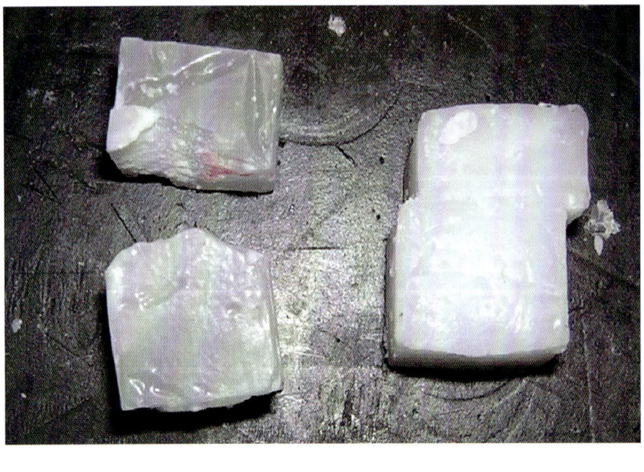

Figure 2
Inherent shear failures of candle wax; the wax at the left exhibits a brittle failure mode, and the wax at the right exhibits a ductile failure mode. Photo: Authors.

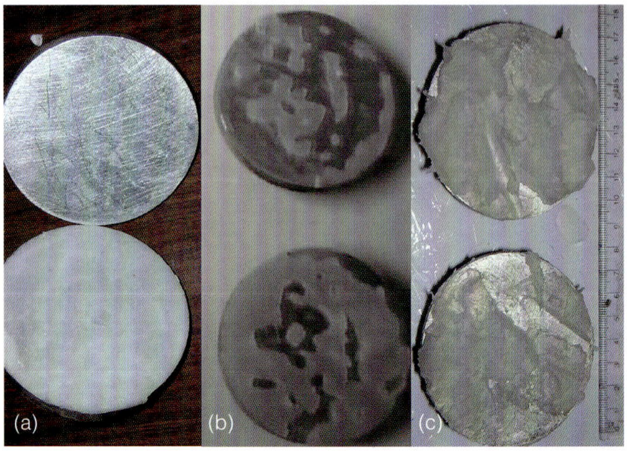

Figures 3a–c
Tensile failures of candle wax; these were categorized as failure of adhesion (a), mixed failure—that is, adhesion failure and cohesion failure combined (b), and cohesion failure (c). Photo: Authors.

Wax	Tensile tests (kPa)				Interface shear tests at 50 kPa (kPa)			
	Hot-applied range		Cold-applied range		Hot-applied range		Cold-applied range	
	No coating	Coating	No coating	Coating	No coating	Coating	No coating	Coating
Multiwax	161–172	254–335	117–208	158–201	45–91	50–149	10–38	15–95
Quake Hold	129–229	209–395	107–170	125–259	7–29	12–37	17–25	43–80
Secure Wax	179–277	293–350	76–123	112–140	17–23	48–56	10–13	23–42

Table 1
Summary of test results.

Capacities for interface shear ranged from 7–91 kPa when hot applied and 10–38 kPa when cold applied. Coating increased cold-applied capacity by 80% but gave less consistent results for the hot-applied samples. Microcrystalline waxes generated an average of a 126% increase in shear strength with a 470% increase in normal load, an approximately 25% increase in capacity as a function of additional normal load. Average inherent shear capacities were over 300% higher than their average interface shear capacities (fig. 4a) and proved a poor indicator of interface shear capacity, despite similar failure mechanisms (fig. 4b).

Analytical Results

SEM scans revealed the waxes' microstructures, which can impact tensile and shear capacities. Paraffin has a linear crystalline structure (fig. 5a). As force is applied, the chains can slide away from one another relatively easily, thereby causing sudden failure, in a brittle manner. In contrast, the branching of the microcrystalline waxes (figs. 5b and 5c) generates a slower, more progressive failure, as the branches act as mini-shear keys along the failure plane, resulting in a ductile failure mode.

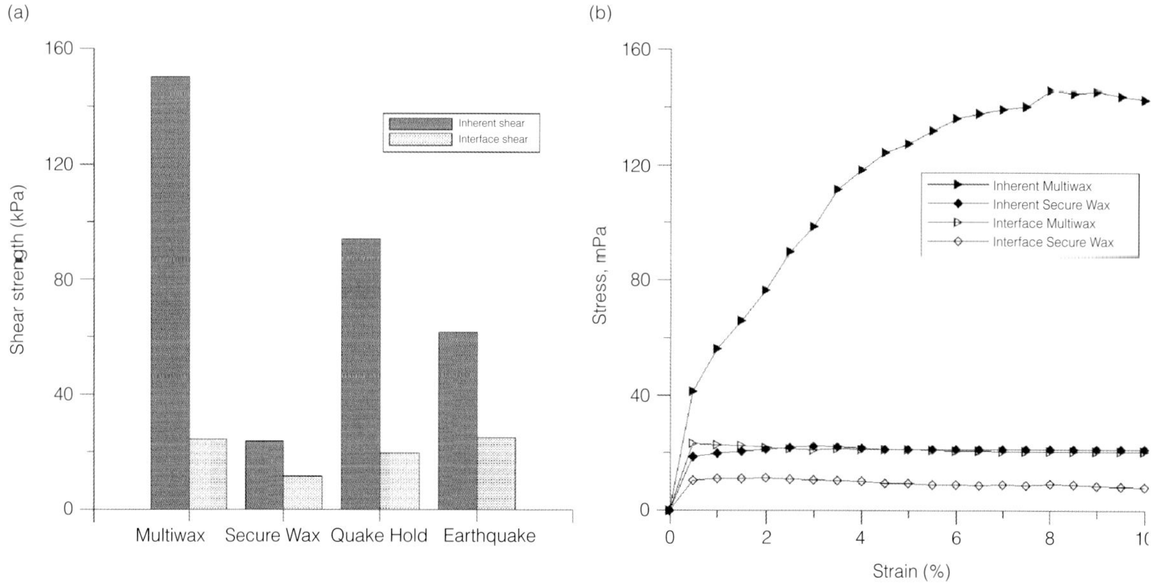

Figures 4a and 4b
Inherent versus interface failure results. Shown are the average results for five tests (a) and the average comparative failure patterns (b).

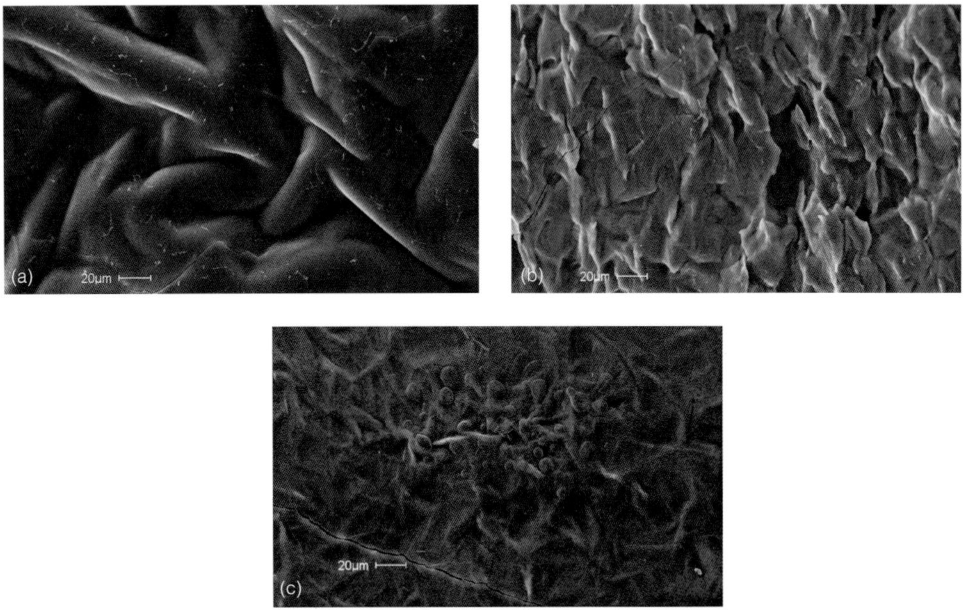

Figures 5a–c
SEM scans of the cast surfaces of the waxes: paraffin (a), Multiwax (b), and Secure Wax (c). Photo: Authors.

Test	Application	Needle penetration	Contraction	Density	Softening/melting point	# Branches
Tensile	Hot with coating			0.99		
	Hot, noncoating			1.05		
	Cold with coating	0.87				
	Cold, noncoating	0.82	1.31			
Shear	Hot with coating					2.03
	Hot, noncoating	1.24				
	Cold with coating			0.98	1.02	0.88
	Cold, noncoating	0.92	1.39			

Table 2
Usefulness of simple, alternative testing methods as a means of predicting tensile and shear tests, as expressed as a percentage of predictive capability.

Simple Measures

Different simplified testing techniques proved effective at predicting specific application arrangements (table 2). Values closer to 1.0 more accurately predicted capacity.

Conclusions and Recommendations

- When applied cold, Multiwax gave consistently high tensile and shear capacity, but to be applied cold, it had to be taken in small pieces (approximately 2 g, or 0.007 oz.) and warmed between index finger and thumb to obtain good adhesion, and it required a large force (approximately 235 kPa) during placement to displace air voids.
- For successful cold wax application, surfaces must be free of contaminants and air pockets.
- Simplified experimental procedures can be used to predict performance but must be selected for a specific application method.

Bibliography

Agrawell, K. M., and G. C. Joshi. "Composition and Properties of Some Microcrystalline Waxes," *Research and Industry* 30, no. 1 (1985), pp. 36–39.

———. "Microcrystalline Waxes—Chain Branching and Its Influence on Physical Properties of Waxes," *Erdol & Kohle Erdgas Petrochemie* 36, no. 6 (1983), pp. 274–76.

———. "Microcrystalline Waxes 1: Investigations on the Structure of Waxes by Proton Nuclear Magnetic-Resonance Spectroscopy," *Journal of Chemical Technology and Biotechnology* 31, no. 11 (1981), pp. 693–96.

American Society for Testing and Materials [ASTM]. "Standard Practice for Verification of Specimen Alignment under Tensile Loading," E 1012 (Philadelphia, 1999).

———. "Standard Test Method for Direct Shear Tests of Soils under Consolidated Drained Conditions," D 3080 (Philadelphia, 2004).

Benuska, Lee, ed. "Architecture, Building Contents and Building Systems," *Earthquake Spectra*, supplement to vol. 6 (1990), pp. 339–76.

Billmeyer, F. W. *Polymer Science* (New York, 1984).

Braun, S., H. O. Kalinowski, and S. Berger. *150 and More Basic NMR Experiments: A Practical Course* (Weinheim, Germany, 1998).

British Standards Institution [BS]. *Adhesives. Determination of the Softening Point of Thermoplastic Adhesives (Ring and Ball), EN 1238* (London, 1999).

———. *Metallic Materials—Tensile Testing—Part 1: Method of Test at Ambient Temperature, EN 10002-1* (London, 2001).

———. *Method for the Determination of Melting Point and/or Melting Range, BS 4634* (London, 1970).

———. *Methods of Test for Petroleum and Its Products, BS 2000-49: Bitumen and Bituminous Binders, Determination of Needle Penetration, EN 1426* (London, 1999).

———. *Methods of Test for Soils for Civil Engineering Purposes—Part 7: Shear Strength Tests (Total Stress), BS 1377* (London, 1990).

Cornu, E., and L. Bone. "Seismic Disaster Planning: Preventive Measures Make a Difference," *Western Association for Art Conservation Newsletter* 13, no. 3 (1991), pp. 13–19.

Crowley, A., D. F. Laefer, and M. Fanning. "Sample Preparation and Testing Methods for the Evaluation of Microcrystalline Waxes for the Seismic Protection of Art Objects," *Testing and Evaluation Journal* [ASTM] 36, no. 1 (2008), p. 10.

Federal Emergency Management Agency [FEMA]. *Reducing the Risks of Non-Structural Earthquake Damage: A Practical Guide* (Washington DC, 1994), p. 9.

Ginell, W. "Seismic Damage Mitigation Techniques for Objects," *Conservation, The Getty Conservation Institute Newsletter* 10, no. 3 (fall 1995), http://www.getty.edu/conservation/publications/newsletters/10_3/feature1_12.html (accessed June 11, 2007).

Goldstein, J. I., et al. *Scanning Electron Microscopy and X-ray Microanalysis: A Text for Biologists, Materials Scientists, and Geologists* (New York, 1992).

Harold, J. M. "Disaster Mitigation for the Bertrand Collection Artifacts," *Cultural Resource Management* 18, no. 6 (1995), pp. 15–17.

Hascall, J. "Artists' Quake Aid Final Report," sect. 10, "Studio Preparedness Earthquake Tip Sheet," http://www.artisttrust.org/about_us/press_room/ARCHIVE/quake_aid/sectionX (accessed April 24, 2007).

Hunter, E. *Practical Electron Microscopy* (Cambridge, 1993).

Kinloch, A. J. *Adhesion and Adhesives, Science and Technology* (New York, 1987).

Koob, S. P. "The Use of Paraloid B-72 as an Adhesive: Its Application for Archaeological Ceramics and Other Materials," *Studies in Conservation* 31, no. 1 (1986), pp. 7–14.

Lever, Y. "Thinking Ahead: Disaster Planning for Museums," *CultureWork* 4, no. 2 (2000), http://aad.uoregon.edu/culturework (accessed April 24, 2007).

Mansoori, A., L. H. Barnes, and G. M. Webster. "Petroleum Waxes," chap. 19 of *Fuels and Lubricants Handbook: ASTM Manual Series: MNL37WCD*, ed. G. E. Totten (West Conshohocken, Pa., 2003), pp. 525–31.

Mashlakian, M. "Earthquake Protection of Museum Artifacts for the Egyptian Art Museum," San Francisco State University, n.d.

Podany, J. C. "Developing and Practicing an Emergency and Disaster Response Plan," *Risk Preparedness for Cultural Properties: Development of Guidelines for Emergency Response* (Tokyo, 1999), pp. 347–59.

———. "The Protection of Cultural Heritage in Museum Collections from Earthquake Damage," *The Symposium on the Conservation and Preservation of the Cultural Artifacts* (Taipei, 1995).

———. "Reassembly and Protection of Large Scale Sculpture at the J. Paul Getty Museum: New Application of Engineering Techniques," *Proceedings of the Sixth International Congress on Deterioration and Conservation of Stone* (Toruń, Poland, 1988), pp. 663–72.

———. "Safeguarding a Collection and Building from Natural Hazards," *Perspectives on Natural Disaster Mitigation* (Washington DC, 1991), pp. 51–68.

Podany, J. C., K. M. Garland, W. R. Freeman, and J. Rogers. "Paraloid B-72 as a Structural Adhesive and as a Barrier within Structural Adhesive Bonds: Evaluations of Strength and Reversibility," *Journal of the American Institute for Conservation* 40, no. 1 (2001), pp. 15–33.

Podany, J. C., and P. T. Leavengood. "Chinese Monumental Sculpture: Treatment and Seismic Stabilization," *Proceedings of the Seventh International Congress on Deterioration and Conservation of Stone* (Lisbon, 1992), pp. 1517–26.

Ratnasamy, P., K. S. Anand, and D. C. Gupta. "Structure and Properties of Microcrystalline Waxes," *Journal of Applied Chemistry and Biotechnology* 23 (1973), pp. 183–87.

Roberts, J. D. *Nuclear Magnetic Resonance, Applications to Organic Chemistry* (New York, 1959).

Russell, T. P., and H. C. Kim. "Tack—a Sticky Subject," *Science Magazine* 285, no. 5431 (1999), pp. 1219–20.

Treloar, L. R. *The Physics of Rubber Elasticity* (London, 1958).

Sources of Materials

Acryloid B-72, adhesives, and consolidants: http://nautarch.tamu.edu/class/anth605/File2.htm (accessed April 25, 2007).

Adhesives and consolidants and archival storage and display: http://www.silcom.com/~css/.

Quake Hold products, waxes, conservation tools, equipment, and supplies: http://www.conservationresources.com/Main/section_39/section39_07.htm (accessed April 25, 2007).

Case Study: Low-Budget Seismic Mitigation Methods for a Museum in Ithaca

Victoria McGuinness

Introduction

Ithaca is one of the smallest Greek Ionian Islands (fig. 1), with a current population of around three thousand, but the archaeological history is both famous and plentiful. Given that Homeric poetry refers to Ithaca when describing Odysseus's difficult journey home, archaeologists have tried to find the fact behind the stories. Schliemann himself excavated on Ithaca, prior to leaving the island in search of Troy.

The Stavros Museum on Ithaca has collections dating from the early Helladic period (3000–2000 B.C.) until the Hellenistic period. The displays include late Mycenaean *kylikes* (sing., *kylix*), clay vessels with a small base, long stem, and wide brim which are shown in an exhibition case in the Stavros Museum (fig. 2). This shape is quite unstable and overturns even during small earthquakes. The kylikes in the Stavros Museum were badly affected in 2003 during an earthquake that measured 6.4 on the Richter scale which originated off the coast of neighboring Lefkás. All of the vessels fell from their display, and 50 percent were badly fractured. With the permission of the Stavros Valley Project, the damaged Kylix 224 was used in this case study (fig. 3).

Purpose of the Study

The overall purpose of this study was to design and manufacture inexpensive seismic mitigation methods for museums that are located in seismic zones and have particularly low budgets. Kylix 224 was used in the research as representative of one of the more difficult object shapes to protect. A main criterion established for the mitigation measures was that the fixtures or mounts designed to stabilize the objects had to be made of inexpensive materials that were also readily available. The mounts also had to be effective—that is to say, they had to be sufficiently strong, durable, and stable and could not be the source of additional stress, load, or damage to the object. The mounts also were to be aesthetically nonobtrusive. With regard to the kylix, the criteria included a need for the mount to be transparent so that it would not obscure any part of the vase.

Results

Thorough testing of the first two mounts produced (fig. 4) led to the modifications that were incorporated into the next set of experiments. The combination of research and practical assessment enabled a thorough and reliable investigation of the suitability of the proposed designs.

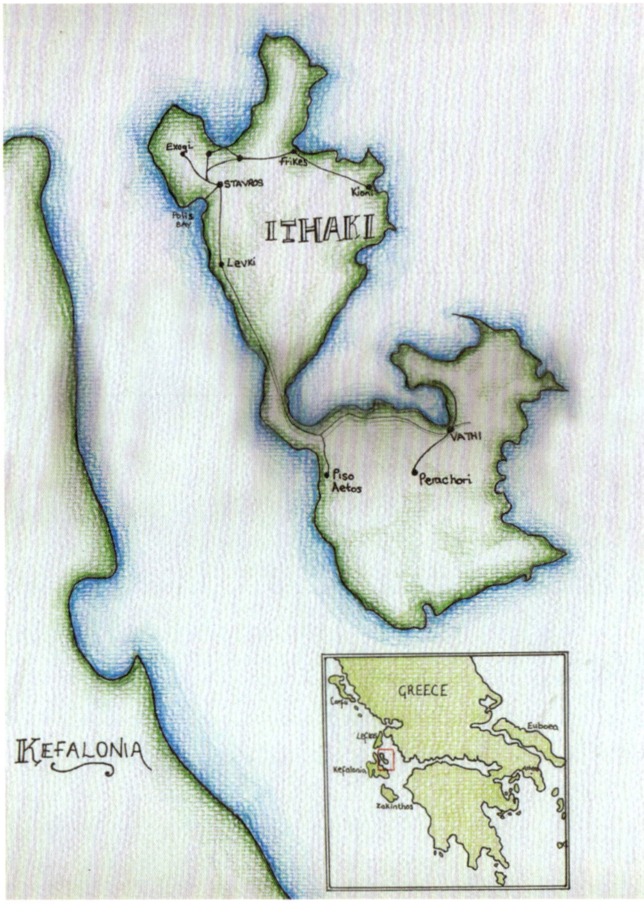

Figure 1
The island of Ithaca. Drawing: Victoria McGuinness.

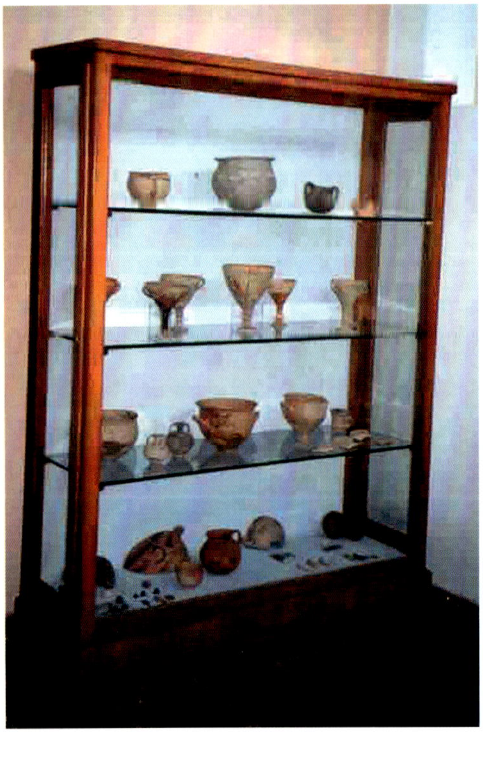

Figure 2
Kylikes displayed in the Stavros Museum. Photo: Victoria McGuinness.

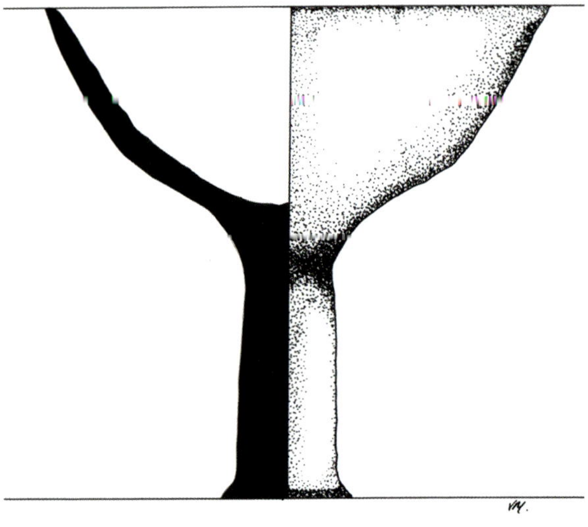

Figure 3
Kylix 224, a late Mycenaean Greek clay drinking vessel, H: 16.4 cm (6½ in.); rim diam.: 12.3 cm (4⅞ in.). Drawing: Victoria McGuinness.

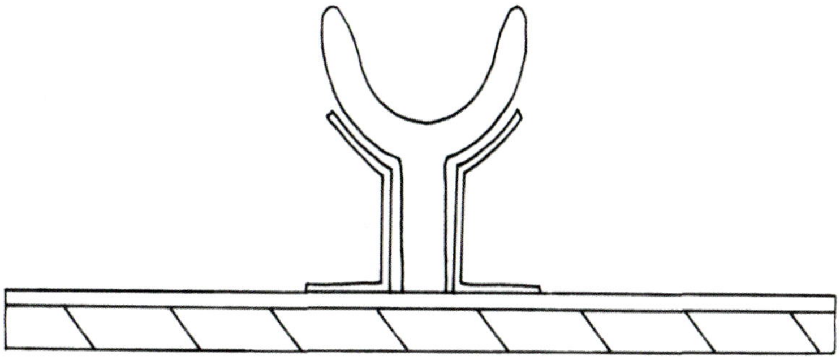

Figure 4
One of the early mount designs. Photo: Victoria McGuinness.

Modifications included wider areas of support around the kylix (fig. 5). Because the support mirrored the kylix's straight edges, the range of vibrations was lessened, and mechanical damage was reduced. Tests at varying frequencies were carried out on each mount design, and measurements were taken of the displacement of the kylix, indicating the potential for mechanical damage. The designs that had edges contoured to the kylix and minimal low-density padding in order to prevent surface damage were the most effective in the experiments.

Mount 4 had the best results at a higher frequency (1.5 Hz), compared to the earlier designs at 1.0 Hz, thus proving that the final design (mount 4) dissipates the effects of earthquakes effectively. The mounts were tested for 30 sec. in the initial experiments; however, the final design was tested for 3 min., in order to impose the high-frequency vibrations for an extended period. The risk of the kylix toppling is minimized when the design incorporates three transparent arms. However, it should be established whether, during an earthquake, the object would damage the mount and, consequently, itself be damaged. At maximum frequency testing (3.0 Hz), the kylix did not cause any damage to the mount. This finding should be taken into consideration when this material is considered for heavier objects.

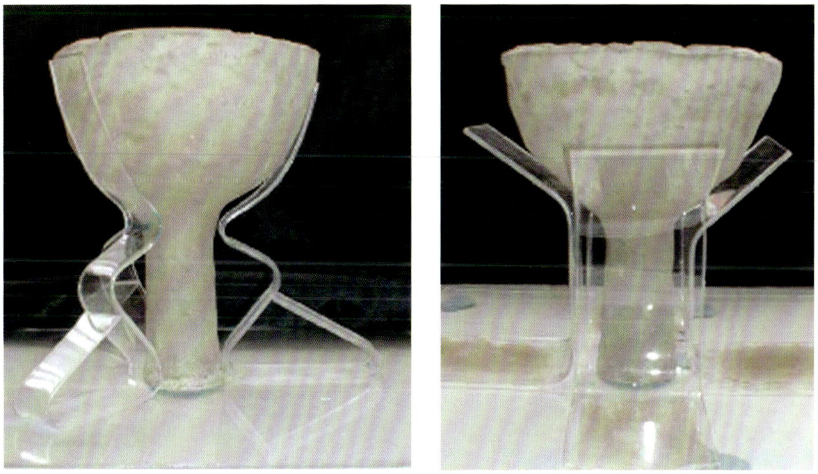

Mount 1

Mount 2

Figure 5
Later modifications of the mount design. Photos: Victoria McGuinness.

Methodology

Research incorporating structural engineering principles resulted in a number of preliminary designs. Of those designs the most suitable were selected for more intensive testing. It was ultimately concluded that acrylic was the most suitable material because of its durability, transparency, and stability.

With the generous support of the University of East London's Civil Engineering Department, the designs were tested on their shake table (Quanser Consulting Inc.) (fig. 6). One of the two accelerometers attached to the test

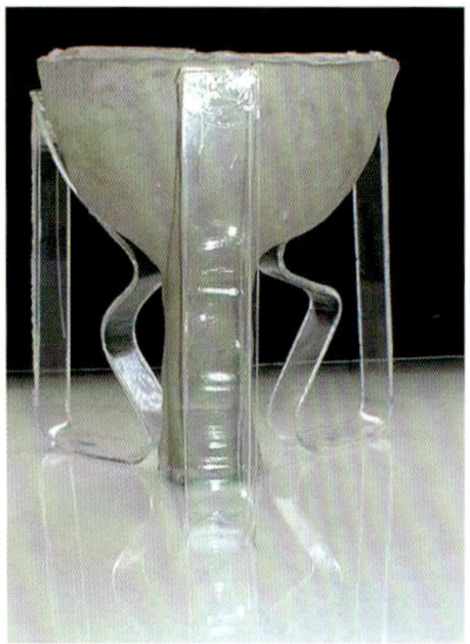
Mount 3

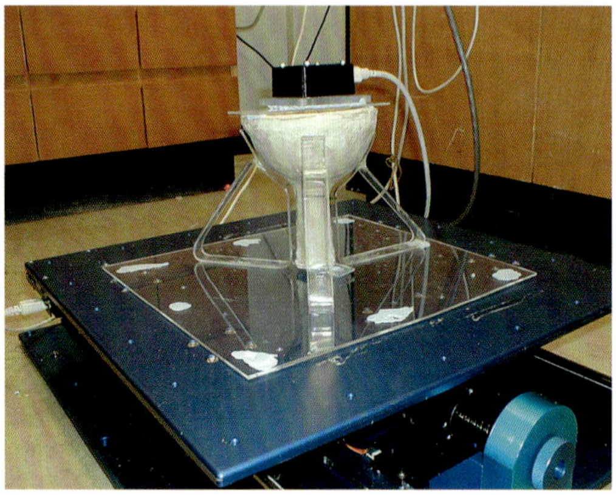
Mount 4

Figure 6
Testing on a unidirectional shake table. Photos: Victoria McGuinness.

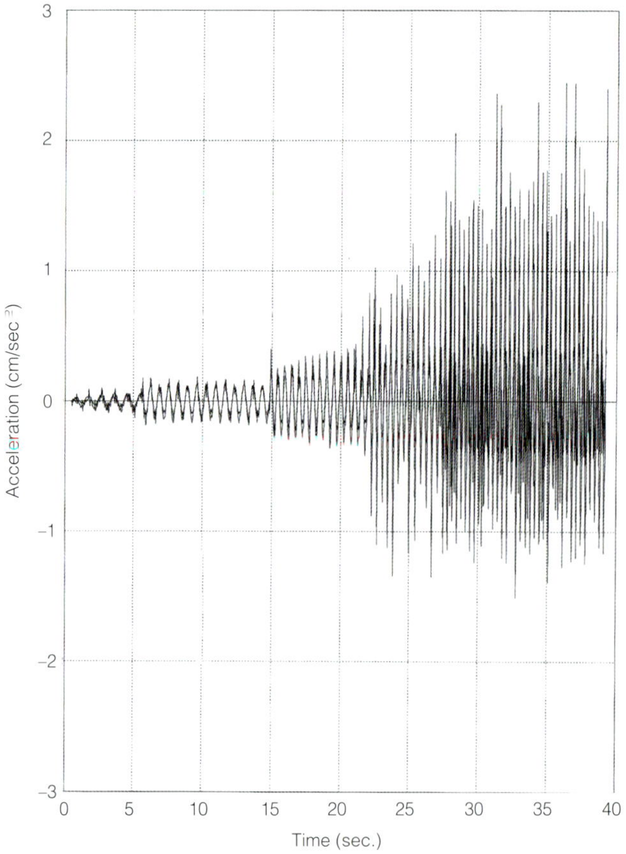

Figure 7
Accelerogram data from the tests.

assembly was always mounted on the shake table, and the second was attached to the kylix in order to record its movements and, consequently, indicate any possible physical damage. Results were displayed on acceleration-time graphs (fig. 7).

Mount 4 matched the specifications of the design brief and efficiently diffused the vibrations, preventing Kylix 224 from toppling and breaking. The strength and durability of mount 4 was tested up to a frequency of 3.0 Hz, during which the kylix neither toppled nor suffered damage.

Future Developments

This research will be developed into a self-help assessment and construction package for small and low-budget museums in seismic zones. Under the aegis of the British School at Athens, seismic mitigation efforts in the current Curation and Conservation Project at the Stratigraphical Museum in Knossos have also been undertaken.

About the Contributors

O. P. Agrawal is director general of the Indian Council of Conservation Institutes, Indian National Trust for Art and Cultural Heritage, New Delhi.

David Armendariz is an assistant mount maker at the J. Paul Getty Museum at the Villa.

Brian Considine is senior conservator of Decorative Arts and Sculpture at the J. Paul Getty Museum, Los Angeles.

Anne Crowley is a PhD candidate at University College, Dublin.

Nicholas Dorman is chief conservator at the Seattle Art Museum.

Ziyad Durón is professor of structural engineering at Harvey Mudd College in Claremont, California.

Eser Durukal works in the Department of Earthquake Engineering at Bogazici University, Istanbul.

Mustafa Erdik works in the Department of Earthquake Engineering at Bogazici University, Istanbul.

Nevra Erturk is a research assistant at Yildiz Technical University, Istanbul.

B. J. Farrar is a mount maker at the J. Paul Getty Museum at the Villa.

William S. Ginell is a scientific consultant at the Getty Conservation Institute, Los Angeles.

Robert W. Graves is principal seismologist at URS Corporation, Pasadena, California.

Karen Hildebrand is a historical archaeologist in the California State Parks.

John W. Hirx is head objects conservator at the Los Angeles County Museum of Art.

Kimio Kawaguchi is conservator of the National Museum of Western Art, Tokyo.

Charles A. Kircher is principal of Kircher and Associates, Palo Alto, California.

Vlasis Koumousis works at the Institute of Structural Analysis and Aseismic Research at the National Technical University of Athens, Greece.

Ioannis Koutromanos is a graduate student in the Civil Engineering Department at the University of California, San Diego.

Debra F. Laefer is director of conservation research at the School of Architecture, Landscape, and Civil Engineering at University College, Dublin.

McKenzie Lowry is a mount maker at the J. Paul Getty Museum at the Villa.

Charilaos Maniatakis is a graduate student in the Civil Engineering Department, Laboratory for Earthquake Engineering, at the National Technical University of Athens, Greece.

Victoria McGuinness is former curation project manager of the Curation and Conservation Project at the Stratigraphical Museum, Knossos, Crete. She currently works at the Ashmolean Museum, Oxford.

Nancy Mendez is an interpretive planner in the California State Parks.

Jerry Podany is senior conservator of Antiquities at the J. Paul Getty Museum at the Villa.

Matthew J. Schoettler is a PhD candidate in the Department of Structural Engineering at the University of California, San Diego.

Robert Sieger is a mount maker with Fine Art Service in Madison, Wisconsin.

Paul G. Somerville is principal seismologist at URS Corporation, Pasadena, California.

Andreas Stavridis is a PhD candidate in the Department of Structural Engineering at the University of California, San Diego.

Donna Strahan is conservator at the Metropolitan Museum of Art, New York.

Constantine Spyrakos is an associate professor in the Civil Engineering Department, Laboratory for Earthquake Engineering, National Technical University of Athens, Greece.

Alex G. Tarics is a seismic engineer who lives in Belvedere, California.

Index

Note: Page numbers in *italics* refer to illustrations.

Acknowledgments

Bringing the work of several professions together so that their combined efforts might contribute to resolving pressing problems is common in our modern world. But when those professions seem to use completely different vocabularies and have what at first appears to be quite diverse working principles, the task of building metaphorical bridges can seem as complex as engineering an actual bridge. That is, unless creative and dedicated people shape the various contributions into a single, shared language. It is these people who are recognized here.

The international conference *Protecting Museum Collections from Earthquake Damage*, held at the Getty Museum on May 3–4, 2006, from which the papers in this volume came, was largely made possible by the support and foresight of Dr. Michael Brand, director of the Getty Museum. His commitment to building, sharing, and protecting the collections of the Getty provided the opportunity for such a broad scope of professionals to combine their efforts and advance the science of preservation.

Thanks go to my friend and colleague Bruce Metro, head of the Preparations Department at the Getty Museum, whose contributions to collections care have spanned three decades of inventive and committed service; to Brian Considine, head of the Museum's Decorative Arts and Sculpture Conservation Department; and to Stephanie Prabulos, Lisa Guzzetta, and Ivy Okamura, who all gave freely of their time and experience in the planning of the conference.

The continuing support of each member of the Department of Antiquities Conservation has been both an inspiration and a willing vehicle for sharing seismic mitigation approaches and developments among colleagues internationally. The 2006 congress speakers, representing six countries facing the threats of seismic activity, are at the heart of this volume. The participants' professional acclaim spans the fields of seismology, seismic engineering, structural and civil engineering, conservation, and collections care.

An enormous debt is owed to Sylvia Lord for her endless patience, insistent perfection, and always ready humor for shaping this volume into something both compact and comprehensible. In Getty Publications, thanks to Mollie Holtman for her editorial wisdom and skill; to Deenie Yudell and Gary Hespenheide for their designers' eye; to Elizabeth Zozom for her work as production coordinator; and to Mark Greenberg for his guidance and patience.

And finally, a well-deserved thank you to the mount makers of the Department of Antiquities Conservation, McKenzie Lowry, B. J. Farrar, and David Armendariz. Their combined abilities, dedication, and creativity bridge the museum world with the world of engineering by dazzling both sides with insights and direct solutions. Although working in the *background*, they form the very front lines of preservation and preventive conservation.

Jerry Podany
Head of Antiquities Conservation
J. Paul Getty Museum at the Villa